Pesticide Drift and the Pursuit of Environmental Justice

Food, Health, and the Environment
Series Editor: Robert Gottlieb, Henry R. Luce Professor of Urban and Environmental Policy, Occidental College

Keith Douglass Warner, *Agroecology in Action: Extending Alternative Agriculture through Social Networks*

Christopher M. Bacon, V. Ernesto Méndez, Stephen R. Gliessman, David Goodman, and Jonathan A. Fox, eds., *Confronting the Coffee Crisis: Fair Trade, Sustainable Livelihoods, and Ecosystems in Mexico and Central America*

Thomas A. Lyson, G. W. Stevenson, and Rick Welsh, eds., *Food and the Mid-Level Farm: Renewing an Agriculture of the Middle*

Jennifer Clapp and Doris Fuchs, eds., *Corporate Power in Global Agri-food Governance*

Robert Gottlieb and Anupama Joshi, *Food Justice*

Jill Lindsey Harrison, *Pesticide Drift and the Pursuit of Environmental Justice*

Pesticide Drift and the Pursuit of Environmental Justice

Jill Lindsey Harrison

The MIT Press
Cambridge, Massachusetts
London, England

© 2011 Massachusetts Institute of Technology

For information about special quantity discounts, please email <special_sales@mitpress.mit.edu>

This book was set in Sabon by Graphic Composition, Inc. Printed and bound in the United States of America.

Library of Congress Cataloging-in-Publication Data

Harrison, Jill Lindsey, 1975–
Pesticide drift and the pursuit of environmental justice / Jill Lindsey Harrison.
   p.   cm.—(Food, health, and the environment)
Includes bibliographical references and index.
ISBN 978-0-262-01598-1 (hardcover : alk. paper)—ISBN 978-0-262-51628-0 (pbk. : alk. paper)
1. Pesticides—Environmental aspects—California. 2. Air—Pollution—California.
3. Environmental justice—California. 4. Spraying and dusting in agriculture—California—Safety measures. I. Title.
TD887.P45H37   2011
363.738'4—dc22

                                                                    2011001921

10 9 8 7 6 5 4 3 2 1

Contents

We are not here to tell you to ban every pesticide, because we are not against ag. We just want to make it better for the people that work around it, live around, or go to school around it.

—Teresa DeAnda, pesticide drift activist in Earlimart, California

It never ends. That's why it's a movement, because it never ends.

—John Mataka, environmental justice activist in Grayson, California

Series Foreword

I am pleased to present the sixth book in the Food, Health, and the Environment series. This series explores the global and local dimensions of food systems, and examines issues of access, justice, and environmental and community well-being. It includes books that focus on the way food is grown, processed, manufactured, distributed, sold, and consumed. Among the matters addressed are what foods are available to communities and individuals, how those foods are obtained, and what health and environmental factors are embedded in food system choices and outcomes. The series not only looks at food security and well-being but also regional, state, national, and international policy decisions as well as economic and cultural forces. Food, Health, and the Environment books provide a window into the public debates, theoretical considerations, and multidisciplinary perspectives that have made food systems and their connections to health and environment important subjects of study.

Robert Gottlieb, Occidental College
Series editor

Preface

The thermometer in my truck read 112 degrees in Delano, California, on the last day that I conducted fieldwork for this book. It felt appropriate—I had spent many sweltering days in the Central Valley. Besides, that day, just like my first one in the field for this project (and many others in between), I was hanging out with pesticide drift activist Teresa DeAnda.

DeAnda lives at the edge of Earlimart, a dusty and tiny farmworking community between Bakersfield and Fresno. She was catapulted into political activism when she and her neighbors were poisoned by an invisible cloud of toxic pesticides that drifted into their neighborhood from a nearby agricultural field in 1999. Because of her many family and work responsibilities, "interviewing" DeAnda typically means chatting with her relatives, riding in the car together while she runs errands, and meeting her neighbors, local officials, and other professional contacts. Persistent yet courteous, she is well respected by seemingly everyone, well connected, and one of the most prominent pesticide activists in the valley.

On this day, DeAnda introduced me to Luis Medellin, one of the youngest people I interviewed for this book. Somehow we had not formally met until that day, although I had seen him around at activist events in the past. We met at the coffee shop in the small town of Lindsay, where he was born and raised. Lindsay is a farmworking community just north of Earlimart, surrounded by orchards filled with oranges, olives, walnuts, pistachios, and stone fruit like peaches, nectarines, and plums. Luis is hefty guy, and he might be intimidating if it weren't for his big smile and professional, positive, and friendly demeanor. His aunt, Irma Medellin, is an environmental justice activist, and he started joining her efforts eight years ago when he was sixteen years old. These days, he is busy managing an air monitoring and biomonitoring project in his town, where residents test their air and bodies for evidence of pesticide drift and exposure to it. After our interview, while he showed me a mural that he helped paint near

Credit: Illustration by Aaron Cole

the center of town, I asked how he stays motivated to devote so much of his personal time to his activist work. He mentioned that a friend of his from high school was diagnosed with leukemia in 2008, and died within a year after the cancer spread from his lymph nodes to his blood, stomach, and brain. Luis does not know whether that illness had any relation to pesticides, but he does know that his own body has been poisoned by pesticides and worries about the effects of those chemicals on his three younger sisters.

My meetings that day confirmed for me that the movement for environmental justice in the Central Valley is in good hands. DeAnda and the Medellins, like so many of the activists I have met over the years, are thoughtful, creative, and optimistic. Their activist work involves educating their neighbors about environmental issues and policy processes, patiently building relationships with allies in local and state government, strategically choosing their battles, and cultivating the next generation of politically active residents. Like DeAnda, Luis Medellin is deeply committed to pesticide regulatory reform and has his heels dug in for the long haul:

We'll never give up. If someone tells you "no," you have got to go back, work on it, and come back until they tell you "yes." . . . If one door closes, another one opens. We have to keep on fighting. Even if it takes six hours, six days, six weeks, six months, six years, sixty years, we're still going to continue the work, no matter how long it takes to win a victory.

I feature the work of pesticide drift activists like DeAnda and the Medellins because I believe that they can inspire us, and also show us how to solve environmental problems more effectively and justly. It is worth pointing out that most pesticide drift activists do not refer to themselves as such, since pesticide drift is only one issue among many that they confront. Water pollution, other forms of air pollution, neighborhood improvements, and school funding are a few of the other main issues that many of them actively tackle. I focus in this book on pesticide drift because, in my view, it provides a compelling lens through which to understand and solve seemingly intractable environmental problems.

I should also warn the reader that I devote a considerable amount of this book to explaining the persistence of pesticide drift as a social problem. I do so because I argue that we cannot solve problems until we really understand where they come from. In particular, I concentrate my analysis on the institutional, cultural, economic, and other structural factors that render so many environmental problems difficult to understand and effectively address.

My overarching goal with this book is to demonstrate that environmental problems are as much about different ideas of what justice means as they are about technical issues or lapses in individual judgment. I want the reader to think about how social inequalities and relations of oppression complicate our abilities to understand and solve environmental problems—and how, in turn, we could use that knowledge to more meaningfully address environmental inequalities. Pesticide drift activists are everyday people who help show us how to do all of these things—how, in short, to pursue environmental justice.

Acknowledgments

This book is my effort to pull apart the multiple notions of justice at work in the world around us, and identify how those ideas then shape the way we solve environmental problems. Melanie DuPuis gets credit for pushing me to let my informants define justice for me, think about the different meanings of justice, and look to political philosophers for help with structuring my arguments. I also gratefully acknowledge Harry Brighouse and the adventurous students in my graduate seminar on political theories of justice at the University of Wisconsin at Madison for graciously helping me figure out how to integrate that immense body of philosophical literature into empirical, critical social science research.

I owe thanks to many people who have helped me understand and write about agriculture and food systems. I am especially grateful to David Goodman, Patricia Allen, Melanie DuPuis, Margaret FitzSimmons, Bill Friedland, Julie Guthman, Andrew Marshall, Dustin Mulvaney, Tim Vos, Keith Warner, and the other members of the Agrifood Studies Research Group at the University of California at Santa Cruz for nurturing critical agrifood scholarship. Sean Swezey taught me an immense amount about integrated pest management, and the Environmental Studies Department at the University of California at Santa Cruz has kindly continued to provide me a soft spot to land every time I venture back to California. Max Boykoff, Sandy Brown, Mark Buckley, Rose Cohen, Brian Gareau, Mike Goodman, Dustin Mulvaney, and many others within and beyond the Political Ecology Working Group supplied friendship and intellectual support—perhaps more than they realize.

I am particularly indebted to pesticide drift activists, who graciously welcomed me into their work and lives, and always had time to meet with me. I hope you feel that I have done justice to your efforts. Susan Kegley deserves special recognition for generously spending many hours helping me understand risk assessment within and beyond the regulatory arena; she is a gold mine of knowledge and patience.

Many representatives of environmental regulatory agencies took time out of their days to meet with me, describing their responsibilities and the constraints within which they work. I recognize that some people working in those institutions may feel that this book is too critical. I believe that most people at the U.S. Environmental Protection Agency and California Environmental Protection Agency are working hard to do their jobs properly and improve the processes of pesticide regulation, and I honor those efforts and the risks that many took to speak candidly with me. Yet I also believe that regulatory institutions are greater than the sum of their parts, and I have a firm conviction that environmental regulatory agencies could do much more to protect public health. I have therefore designed this book to focus the reader's attention on the institutional, cultural, and other structures that uphold pollution and illness despite individual regulators' efforts to address those problems, and I also offer a set of concrete suggestions for making regulation more just. This book is written for regulatory agency scientists and managers as much as activists and academics, and I wrote it in the spirit of making California a better place to be—for everyone.

My colleagues at the University of Wisconsin at Madison helped to make work such a pleasant experience; I thank them for encouraging me at every step and having faith that I could complete the task. Graduate students Sarah Lloyd, Julia McReynolds, and Trish O'Kane did considerable research, and also provided thoughtful reflections on many of these ideas. Special thanks to Phil Brown, Nik Heynen, Hilda Kurtz, Mara Loveman, Geoff Mann, Don Mitchell, Scott Prudham, and Laura Senier for insightful and speedy advice at several crucial points as well as inspiration all along the way. Bob Gottlieb helped to improve this book in no small way, notably for pushing me to think more broadly about alternative agrifood activism. The editorial staff at the MIT Press were a delight to work with and significantly polished this book. Thanks also to the anonymous reviewers for their thoughtful, constructive suggestions.

The research for this book was funded in part by the University of California Institute for Labor and Employment, the University of California at Santa Cruz Graduate Division, the Department of Environmental Studies at the University of California at Santa Cruz, Frederick H. Buttel Professorship Funds at the University of Wisconsin at Madison, a Wisconsin Alumni Research Foundation faculty grant, and the Program on Agricultural Technology Studies at the University of Wisconsin at Madison.

Some of the ideas in chapters 4 and 5 were previously published by Elsevier in "Abandoned Bodies and Spaces of Sacrifice: Pesticide Drift

Activism and the Contestation of Neoliberal Environmental Politics in California," *Geoforum* 29 (2008): 1197–1214, and "'Accidents' and Invisibilities: Scaled Discourse and the Naturalization of Regulatory Neglect in California's Pesticide Drift Conflict," *Political Geography* 25 (2006): 506–529.

Though they may not realize it, my parents had a major influence on this research, since they jump-started my compassion for migrants at an early age. I also thank my parents, Jim, Max, and others for giving me quiet, sunny spaces to write at crucial points along with the encouragement to get it done. I thank Max most of all: You provided endless encouragement and patience, and kept me happy and healthy. I owe you one.

1

Introducing Environmental Justices

In the past forty years, the environmental movement has radically trans-formed how we think about the interrelationships between social and ecological systems. Rachel Carson's *Silent Spring*, published in 1962, was a crucial moment in the rise of environmental politics, putting a trenchant and scientific critique of the disastrous impacts of "modern" chemical technologies into engaging prose that resonated with the general public. The environmental justice (EJ) movement that has emerged and grown since the 1980s has pushed this critique further, arguing that meaningfully confronting environmental problems requires attention to the ways that they shape the lives of some social groups more than others. Levying its charge at the state, industry, and the mainstream environmental move-ment itself, the EJ movement has shown that mainstream environmental politics have typically ignored the fact that the world's most vulnerable and marginalized groups bear a disproportionate share of environmental burdens. EJ activists and scholars explain that various forms of political *injustice*—corporate profit seeking and malfeasance, the state's failure to adequately represent or protect the needs of marginalized social groups, and other forms of "raw power" that have "pitted the powerless against the powerful all over the world"—have forced poor people and people of color to bear a disproportionate share of environmental harms.[1] In other words, it is by disregarding justice that powerful actors are able to shift environmental burdens to the people who are least able to contest them.[2] Accordingly, EJ activists and scholars advocate bringing justice into environmental politics.[3]

My aim in this book is to both uphold and amend this EJ argument. This book pivots around political conflict over pesticide drift in Cali-fornia—a case that illustrates in sharp, present detail how the workings of "raw power" shift the burden of pesticide pollution to the bodies of California's most marginalized and vulnerable residents. That said, I also challenge the claim that environmental inequalities exist because mainstream (i.e., non-EJ) environmental politics are devoid of justice. I contend instead that environmental inequalities emerge from cruelty and malfeasance, but also from the ways in which many well-intentioned actors are engaging in efforts to make California agriculture more environmentally sustainable. I use the case study of pesticide drift to demonstrate that contemporary environmental politics are shaped in part by particular notions of justice. I specify and explain these notions later in this chapter, and throughout the book I identify the various roles that these theories play in environmental politics—in some instances expressly constituting the moral charge for particular programs, in other cases co-opted strategically and incompletely to discursively legitimize other programs, and in still other cases invoked unintentionally. I describe why certain theories of justice and the practices they endorse in mainstream environmental politics do little to effectively address problems like pesticide drift, and I show how pesticide drift activists, like the broader EJ movement, push for a set of solutions based in a different notion of justice. The tension between the EJ movement and mainstream environmental politics, in this light, also can be understood as a clash between competing conceptions of justice. In order to fully appreciate and effectively apply the insights of the EJ movement, we must critically interrogate the conceptions of justice that increasingly pervade mainstream environmental politics today (and to which the EJ movement itself is reacting): how they work, why they are problematic, and why they seem reasonable to so many people.

The Case: Political Conflict over Pesticide Drift in California Agriculture

Without a doubt, pesticide illness constitutes one of the most widespread environmental problems today. The United Nations Environment Program estimates that one to five million pesticide poisonings occur every year worldwide, and twenty thousand of those are fatal.[4] What makes these statistics especially chilling is the fact that they represent only the tip of the iceberg, since they do not account for pesticide-related, delayed-onset diseases, nor the fact that most pesticide exposures are neither recognized, treated, nor reported.[5] As scientific evidence amasses about the uncontrollability of pesticides as well as the issues around their long-range transport, people around the world collectively organize to fight against the most highly toxic pesticides and the ways in which they pollute water, air, and food. Pesticide *drift* is the airborne movement of agricultural pesticides into residential areas, schools, and other spaces, and is now a key target of activists' anger, because the wayward movement of pesticides, often far from where they are applied, reveals just how pervasive and under-recognized pesticide exposures actually are. In recent years, public concern about pesticide drift has generated activist campaigns throughout the world, in both the global North and South. In the United States, activist groups all over the country—including Hawaii, Alaska, Maine, the Southeast, the Midwest, California, Colorado, and the Pacific Northwest—are carrying out this work.

California provides an illuminating window into the problem of pesticide drift and its potential solutions. In many ways, California is similar to agriculture-intensive regions throughout the world. California agricultural pesticide use rates are high, pesticide drift has been well documented there, and human exposures to pesticide drift are a regular feature of its agricultural landscape. Though California agriculture is famous for its "industrial" character—highly mechanized and capital intensive—it is also exceptionally labor intensive and interfaces intimately with residential neighborhoods (include long-standing farm towns, new suburban developments, and nearby urban centers).

What makes California different and thus an unusually interesting case study is the fact that many different actors—from industry, the state, and activist groups—have struggled for years to bring agricultural pesticide problems like drift under control. California is in many ways the vanguard of environmental protections, as its long history of environmental activism, famous national parks and other protected lands, and EJ policies and programs exceed those of most other states in the nation. These progressive environmental politics extend into its agricultural sector, where industry innovations, regulatory leadership, and vibrant agrifood activism set it apart from other states in terms of environmental sustainability efforts. Although California agriculture, a $38 billion powerhouse, has long been recognized as the epitome of modern, industrial agricultural production that pivots around highly toxic chemical pesticides, a wide variety of actors in the agricultural industry have invested in innovative efforts to make California agriculture more environmentally sustainable.[6] California's pesticide regulatory apparatus similarly contributes to such sustainability developments. The California Department of Pesticide Regulation (DPR) and the state's County Agriculture Commissioner offices together comprise the largest pesticide regulatory apparatus in the nation, employing hundreds of scientists, managers, and other staff across the state. Additionally, the state's pesticide laws and regulations exceed federal standards in countless ways, support many innovative programs designed to reduce pesticide risks, and often set the bar for federal environmental policy changes. Environmental sustainability efforts by the agricultural industry and regulatory agencies have developed over time in tandem with tremendous public interest in environmental issues along with a dynamic collection of environmental, labor, and food activists who keep agri-environmental issues on the public and political agendas.[7] Throughout this book, I refer to such activism as the "alternative agrifood movement"; pesticide drift activism overlaps with it, but also differs in several notable ways, as I elaborate in chapter 5.

However, despite the "greening" of food and agriculture in California, large-scale pesticide drift incidents have occurred with disturbing regularity in recent years, frightening and sickening thousands of people near agricultural fields. Therefore, California is the perfect case for asking two important questions: Why does this environmental problem persist despite considerable industry innovation, regulatory action, and public activism? How can such efforts be reformed to better address this and other pressing environmental problems?

As it turns out, it is rather difficult to quantify just how pressing an environmental problem pesticide drift is. Official regulatory data indicate that in an average year, several hundred Californians are made ill by agricultural pesticide drift. Regulatory officials emphasize that these incidents are relatively few in number and assert that they are generally caused by applicator error. Starting in 1999, though, a series of remarkably large-scale pesticide drift incidents in California pushed the issue into the spotlight unlike ever before. Crucially, these incidents helped to mobilize political activists who cast doubt on regulators' claims about the scope of the problem.

An incident in Earlimart in 1999 garnered particular attention. Throughout the course of the evening of November 13, at least 170 residents of the small, agricultural community of Earlimart repeatedly experienced frightening and inexplicable acute illness, including vomiting, impaired breathing, dizziness, and burning eyes and lungs. Emergency crews responding to the scene did not speak Spanish and thus could not effectively communicate with many of the residents. Moreover, they could not identify the cause of the illness and were unsure of how to advise the victims, telling some to stay indoors while directing others to leave the vicinity. Eventually, later that night, emergency crews evacuated some of the most ill residents to a nearby middle school, stripped them in front of their neighbors and television crews, and sprayed them repeatedly with fire hoses. A subsequent investigation revealed that a poisonous cloud of a soil fumigant called metam sodium, a known carcinogen as well as reproductive and developmental toxicant, had volatilized more quickly than anticipated from an agricultural field one quarter of a mile away, drifted into the town, and poisoned the residents. Victims were left with fear, lingering illnesses, and medical bills they could not afford to pay. The Earlimart incident helped to expose the inadequate communication between county agriculture commissioners and emergency responders, prompted an investigation by statewide political and regulatory officials, and inspired numerous residents to form a community-based organization (El Comite para el Bienestar de Earlimart) to confront pesticide drift and other problems in their neighborhood.[8]

As much as the Earlimart incident revealed the dangerously unruly nature of agricultural pesticides and emergency responders' numerous failings, subsequent situations showed that Earlimart was not an anomaly but rather part of a regular trend (see table 1.1 below).[9] In November 2000, at least thirty-five elementary school children and several teachers in Ventura County were taken sick after a cloud of chlorpyrifos drifted into the school grounds from a nearby lemon orchard. Chlorpyrifos is a neurotoxic organophosphate insecticide and has been classified as a suspected endocrine disruptor and possible developmental or reproductive toxicant. Unfolding in a largely white, upper-middle-class coastal community, the Ventura incident illustrated that all residents living near agricultural fields are at risk of exposure to pesticide drift.[10]

Subsequent large-scale incidents continued to push pesticide drift into the spotlight. In July 2002, a wayward cloud of metam sodium drifted into Arvin, a farmworking community in the Central Valley on the outskirts of Bakersfield. Initially, the news reported that only one person had been made ill from exposure to the pesticide drift. Hearing rumors that made them doubt the validity of that number, a group of concerned residents from nearby towns and representatives from a regional EJ organization walked door to door to interview neighbors and collect illness data. Their efforts uncovered a pesticide drift event startling for both its size and relative invisibility. They found that at least 273 people living and working in Arvin had likely been poisoned that day, with one woman hospitalized for a week.[11] The following statement from one of those volunteers, Teresa DeAnda (who herself had been poisoned in the Earlimart incident in 1999), conveys the fear and frustration that many residents experienced:

In 2002, when Arvin happened, we went up there and the news report said that only one person had been taken to the hospital. And we didn't believe—I didn't believe that. And I kept telling the county ag commissioner. He wouldn't go. He

Table 1.1
Selected Major Pesticide Drift Incidents in California, 1998–2007

Date	County	Number of people affected	Pesticide
1998	Monterey	12 workers	Diazinon
1998	Merced	12 workers	Chlorpyrifos
1999	Madera	10 children and bus driver	Chlorpyrifos
November 1999	Tulare (Earlimart)	170 residents	Metam sodium
June 2000	Tulare	24 workers	Chlorpyrifos
November 2000	Ventura	35 children and teachers	Chlorpyrifos
June 2002	Kern (Arvin)	138 workers	Metam sodium
June 2002	Kern (Arvin)	273 workers and residents	Metam sodium
October 2003	Kern (Lamont)	163 residents and 3 workers	Chloropicrin
May 2004	Kern (Arvin)	122 workers	Methamidophos
2004	Monterey	11 workers	Diazinon and mefenoxam
May 2005	Kern (Arvin)	27 workers and 6 emergency crew	Cyfluthrin and spinosad
October 2005	Monterey (Salinas)	324 residents	Chloropicrin
August 2005	Kern	42 workers	Metam sodium
September 2006	Sacramento	48 workers	Disulfoton
2006	Merced	10 residents	Methyl bromide and chloropicrin
2006	San Bernadino	51 residents and workers	Chloropicrin
2007	Monterey	31 residents and workers	Methyl bromide + chloropicrin
July 2007	Tulare	28 workers	Chlorpyrifos

said, "What matters to me is how the accident happened and not how far it got, not how many people were sickened by that." That was crazy. So we went door to door. And the first time we got about 40 people that were affected. Their stories were identical to the Earlimart stories . . . and the next time we went out, [we found] 91 people [who had been poisoned]; and then after the DPR people got involved and the county ag commissioner somewhat got involved. Then it was 268 people that were affected in Arvin from that drift. And by being affected, these people were just inundated with the smell, kids vomiting in the front yards, people coughing. One woman said she felt like she was going to die. She could not even breathe. She said, "I thought the big bomb had attacked." She thought it was a terrorist attack.[12]

One year later, another large-scale pesticide drift incident occurred in Lamont, which like Arvin and Earlimart, is located in the southern end of California's Central Valley. On October 3, at least twenty-four Lamont residents suffered a range of acute toxicity symptoms, including nausea, vomiting, blurred vision, and impaired and painful breathing, after the highly toxic soil fumigant chloropicrin drifted from a field one-quarter of a mile away. Emergency crews responding to the scene determined that the symptoms were not severe or persistent enough to warrant further investigation, and they instructed residents to return home and air out their houses. The second half of the pesticide application proceeded the following day, again drifting into the same residential area, and this time causing illness among over two hundred additional residents. On this second day, the victims were evacuated to a nearby parking lot, where they waited for several hours without food, water, medical treatment, or access to bathrooms.[13] Barricades were set up on the edge of town, and emergency response crews prevented residents from leaving the area. Despite officials' claims to the contrary, the Lamont incident demonstrated that regulatory agencies had made little progress on the issue of pesticide drift—failing to even improve incident response protocol, which is the most basic and reactionary of changes clearly needing to be made.[14]

Like pesticide drift incidents that have occurred elsewhere throughout the United States and around the world in recent years, these and other California ones attracted the media's attention; undermined industry's claims about pesticides as controllable; illustrated the inhumane, incoherent, and ineffective nature of regulatory agencies' incident response protocol; raised questions about the role of race, class, and legal status in shaping pesticide use and regulation; and inspired various residents and other activists to collectively organize in order to take on the problem of pesticide drift. The California residents who participate in such grassroots activism live in agricultural communities across the state and include a diverse array of Latino/a farmworkers and their family members, other low-income agricultural community residents of color, and white, middle-class, and upper-middle-class professionals. Several regional and statewide nongovernment organizations (NGOs) have played an important function in cohering these various grassroots strands of pesticide drift activism and thus are featured prominently in this book. Notably, the San Francisco–based Californians for Pesticide Reform (CPR) organizes all of the disparate groups working on pesticide drift into a statewide coalition, the San Francisco–based Pesticide Action Network (PAN) of North America serves as the scientific arm of the nascent movement, and the Sacramento-based Pesticide Watch has periodically provided organizing assistance to community groups interested in pesticide drift.[15] California Rural Legal Assistance, United Farm Workers (UFW), and the Center on Race, Poverty, and the Environment, all with multiple offices around the state, have also provided crucial institutional support to community-based groups active in pesticide drift politics. These organizations collaborate with other environmental organizations throughout the United States to share strategies and resources, and partner on national-level and international campaigns.

While many of these residents and other activists initially started politically engaging in pesticide drift in response to one or more large-scale incidents, their continued commitment to the issue stems from a shared conviction that pesticide drift is a part of everyday life, contributes to an endless array of health problems, and is largely ignored by regulatory officials. In other words, as egregious as the big incidents are, activists view them as unfolding on a landscape of less dramatic but pervasive agricultural chemical contamination and regulatory neglect. Activists' stories, the ways in which they conflict with those of regulatory officials and industry, and their tremendous implications for environmental regulation and public health fueled my own interest in the subject, and these tensions constitute the heart of this book.

The ongoing nature of pesticide drift despite efforts by the agricultural industry, environmental regulatory agencies, and alternative agrifood movement to make agriculture more environmentally sustainable—as well as the conflicting stories told about the problem—raise fundamental questions that must be examined to understand this environmental problem and its solutions. Why do pesticide drift incidents occur in a context of progressive environmental change? How do we explain the coexistence of two completely different interpretations of the problem itself? Which of these has guided the regulatory response to pesticide drift, and with what consequences? Like other scholars of EJ, I argue that understanding these contradictions requires that we recognize pesticide drift as not only a technical problem but also a social one, rooted in systems of inequality and oppression. Moreover, I emphasize throughout this book that environmental inequalities today must also be understood within a broader context of mainstream environmental politics dominated by particular— and particularly problematic—conceptions of justice.

A Technological and Social Problem

In some ways, pesticide drift is a complex, technical problem best understood by medical and environmental scientists. First, the study of pesticide drift includes analyzing the countless ways in which pesticides move through, change in, and interact with the environment. The nine-hundred-plus pesticide active ingredients registered for use in California are manufactured into over thirteen thousand different formulations, in which various amounts of different pesticides are mixed together and applied with innumerable "inert" ingredients that help the pesticide reach and/or adhere to its target.[16] All of these various formulations interact with each other and the ever-changing environments into which they are applied in countless ways, most of which are poorly understood. Also, pesticide drift analysis includes studying pesticide exposure, such as the various pesticides' different routes of exposure (dermal, dietary, or inhalation) and the extent to which some human populations (especially children and farmworkers) are subject to higher rates of exposure. Finally, analysts must take into account the actual health effects of exposure to the various pesticides, where every pesticide interacts with the human body in its own way, produces or contributes to its own collection of health problems, interacts in unknown synergistic or cumulative ways with other environmental toxins, and affects certain sensitive populations (children, fetuses, the elderly, the ill, and the chemically sensitive) more than the "average" body. I elaborate on these technical complexities in chapter 2 of this book.

That said, pesticide drift must be understood as a social problem as much as a technical one, and the intersections between these social and technical dimensions explain the continuation and invisibility of pesticide drift. As I will illustrate throughout the book, experts' abilities to understand and control pesticide drift are challenged not simply by the technical complexity of agricultural pesticides but also from the highly unequal and oppressive social relations in which they are used. Although pesticide drift affects all people living in and near agricultural fields, farmworkers and their families are exposed most frequently. I will show how the poverty, legal status issues, language barriers, political disenfranchisement, and other forms of social marginalization widespread in farmworking communities tend to obscure pesticide exposures and other problems. I will show as well how other pesticide drift victims and activists, although more empowered than immigrant farmworkers, are nonetheless marginalized within the environmental regulatory arena and by mainstream pesticide activism. At the same time, various industry groups exert extraordinary influence within environmental regulatory and policy institutions. Industry groups' financial power, strong coherence, scientific resources, and social networks enable them to shape the terms of regulatory debate in ways that residents of agricultural communities are simply unable to do. Environmental regulation consequently has been bounded by a narrow interpretation of pesticide drift as a series of isolated, unfortunate events requiring minimal regulatory change.

Justice in Environmentalism

It is because of these social factors that pesticide drift can be conceptualized as an *EJ* problem. Since at least the 1980s, the EJ movement has made a scathing critique of the environmental regulatory state and mainstream environmental movement alike for being inattentive to the uneven distribution of environmental problems as well as the ways in which social inequalities inhibit environmental problem solving. The EJ movement is actually a diverse collection of activist groups that primarily represent a confluence of antitoxics activism (with its economic analysis of corporate power and economic structures of pollution) and civil rights activism (with its critique of social structures of race-based oppression).[17] EJ groups loosely align along a common framing—namely, that the distribution of environmental problems is inextricably linked with poverty, racism, and other forms of oppression, and that these same social factors unfairly shape the ways in which the environmental regulatory state interprets and addresses environmental problems. EJ activists also levy their critique at the mainstream environmental movement, arguing that the latter has ignored and thus reproduced environmental inequalities by focusing on protecting wilderness and endangered species, sidelining the environmental issues facing poor communities and communities of color, otherwise privileging a conception of the environment dislocated from relations of social inequality, and relying on litigation, legislation, and other pathways to environmental change that exclude so-called nonexperts from participation.[18]

The EJ movement explicitly conceptualizes environmental problem solving as a question of justice, and accordingly, is widely understood as innovative in bringing *justice* into the conversation of environmental politics. In common language, people refer to justice as a singular concept in this way—as the epitome of fairness, or some unquestionable *right* state. People lament the absence of justice throughout much of the world, and many academics and other critical writers who study social problems call for greater attention to social justice in their various fields and disciplines. Yet justice is not an uncontested concept. In fact, Western philosophers have long debated the meaning of political justice, opening that black box to rigorous interrogation. Several prominent scholars have explicitly articulated the specific and multiple conceptions of justice advocated by the EJ movement; the work of Luke Cole and Sheila Foster, David Pellow, David Schlosberg, Iris Young, Christian Hunold, and Robert Figueroa have particularly influenced my own work in this regard.

My goal in this book is to push this line of inquiry a bit further. I contend that it is neither accurate nor useful to think about the world around us—the one that the EJ movement actively confronts—as generally devoid of justice. Whereas the EJ movement is typically framed as being unique in its concern for justice or fairness within environmental politics, I aim to show that it is more accurate and instructive to conceptualize the EJ movement's claims to justice as a reaction to other, more prevalent notions of justice that deeply and widely shape and are upheld by mainstream environmental institutions and practices. To do this, I draw on the work of a handful of scholars—notably, Melanie DuPuis, David Harvey, and Iris Young—who identify the political theories of justice that shape not just EJ activists' but also more "mainstream" actors' approaches to solving environmental problems. Although we live in a world that is far from ideal or just, certain theories of justice nonetheless shape the design of political institutions and policies. As will become clear, environmental inequalities stem not only from a lack of knowledge, care, or political will but also from many actors' attempts to do the right thing.

Every theory of justice specifies its own vision of a just or fair society, and as such, the appropriate responsibilities of the state vis-à-vis the economy and the public in regard to questions of freedom, equality, participation, and other sorts of political rights. In this book, I elaborate on the theories of justice that define the context in which EJ and other social movements unfold. To suggest that justice was absent from environmental politics before EJ came along would eliminate the important opportunity to highlight the dominant theories of justice that undergird mainstream efforts to solve environmental problems, how those theories of justice are socially and environmentally problematic, how and why their influence has increased over time, why they seem so natural and reasonable, and why effective and fair environmental problem solving requires a different notion of justice itself. Predominant ideologies—including particular conceptions of justice and fairness—are important to understand, since they reinforce and legitimize capitalist expansion, the weakening of the environmental regulatory state, and the associated environmental fallout. These predominant notions of justice must be spelled out, and their material forms critically examined, before we can advocate a different and better vision of justice, and hence a vision of a just environmental politics.

Throughout this book, I will outline the ways in which ideas about justice shape the primary means through which various actors try to address pesticide drift—the agricultural industry (chapter 3), the environmental regulatory state (chapter 4), and alternative agrifood activists (chapter 5). I emphasize how these ideas of justice function *ideologically*—making existing social structures and institutions seem natural and necessary. In contrast, pesticide drift activists, like the EJ movement in general, argue for a different conception of justice, and throughout the book I showcase the ways in which they pursue this goal. Like many other EJ groups, most pesticide drift activists work in grassroots, community-based groups with little or no funding, fight toxics in their neighborhoods, point out and contest the ways in which pollution and illness stem from various forms of oppression (including, but not limited to, those of race and class), and demand entirely different roles for scientific uncertainty and public participation in environmental problem solving. I highlight their work as a grounded, current, compelling story that illustrates the logic behind the EJ movement's claims to justice. That said, pesticide drift activists use EJ framings strategically and irregularly, abandoning them at times to build alliances with other, less radical activist groups. I elaborate on these activist practices, using my observations to discuss both the contributions and limitations of EJ arguments.

Justice in the Literature

This book is thus a story about how environmental problems like pesticide drift continue within a context of increasing environmental activism and the mainstreaming of environmental politics. Empirically, as mentioned earlier, this book concentrates on California agriculture. Scholars have paid considerable critical attention over the years to systems of food production, distribution, and consumption because of their tremendous impacts on the planet and all its inhabitants. Agriculture is a key sector of global and local economies, the direct source of employment and livelihood for hundreds of millions of people, a primary source of open space, the largest use of land on our planet, a tremendous manipulation of natural resources, one of the largest sources of air and water (and now genetic) pollution, the source of human sustenance, and a space in which we negotiate and interrogate our relationship with the natural world as well as each other. Scholarship for academic and laypeople alike provides keen insights along with unique perspectives that help us understand how social inequalities and environmental problems develop in agricultural systems and later become obscured, neglected, and contested.[19]

Recently, such work has come to focus on the politics of "sustainable" agrifood systems—a framework for reforming agrifood systems to more meaningfully incorporate the principles of ecology, economic viability, and social justice.[20] In terms of social justice, most scholars have concentrated exclusively on economic justice for small-scale farmers, and a much smaller but growing body of scholarship is directly addressing food justice issues facing low-income eaters and the labor justice issues experienced by farmworkers. Yet little attention is being paid to social justice as it relates to the environmental context of agricultural pesticide use. Although specifically pertaining to agriculture, this silence points to questions that increasingly structure the work of a broad range of scholars: In what ways do real people actually experience environmental problems and regulations? In what ways do those experiences vary between social groups and across space, and what factors shape that unevenness? Why do some groups seem to be able to influence how a particular environmental issue is regulated, and how do other groups' viewpoints and experiences become marginalized?

Scholarship on EJ activism speaks directly to this silence, paying explicit attention to the ways in which social inequalities exacerbate environmental problems, help to distribute them unevenly, obscure them from public view, and complicate seemingly straightforward solutions. Such research often explores specific case studies of EJ activism, highlighting the efforts of activists who organize themselves in response to local environmental problems and critically confront the multiple forms of injustice that produce and bolster unequal environmental outcomes.[21] Many scholars have stressed the ways in which antitoxics activists and others in the EJ movement have gained traction by mobilizing not just material resources but also nonmaterial, symbolic ones, such as the compelling cognitive "frame" of EJ. The EJ frame has been shown to serve as a crucial mechanism through which residents rally each other and through which disparate community-based battles cohere into a movement, especially where the hazards themselves are scientifically ambiguous.[22]

Research on antitoxics activism helps to politicize the analysis of scientific research, showing how expert systems of knowledge and scientific standards of proof tend to privilege polluters and thus reinforce patterns of illness and the social inequalities they stem from.[23] Moreover, such research illustrates the convictions and insights of grassroots activists, thereby problematizing the standard assumption that formally trained experts are the only bearers of legitimate knowledge. These social movements and the academic analyses about them have also helped to demonstrate that patterns of pollution and illness are deeply rooted not only in malfeasance but also in dominant social ideologies (especially modernist ideas about human dominance over nature, an unfailing optimism about technology, and a belief that increasing production can solve social problems).

Environmental justice researchers also turn their critique to the mainstream environmental movement, arguing that its prioritization of middle-class conceptions of the environment effectively marginalizes the environmental burdens endured by the poor and communities of color. Such analyses draw on a growing body of work in environmental history that critiques unreflexive accounts of environmentalism that privilege conservationism and exclude the role of racism and human labor in "nature."[24] Critiquing the ways in which the mainstream environmental movement and environmental policies have ignored the effects of social inequalities, EJ scholars call for bringing *justice* into environmental politics.[25]

That said, the meaning of justice itself has remained a nebulous and underspecified concept. This has been ameliorated to a considerable extent in recent years as EJ has captured the attention of political philosophers, who draw explicitly on contemporary political theories of justice to clarify the justice claims made by EJ scholars and activists. This turn serves an important function. As Andrew Dobson and other philosophers have observed, the ability of justice to contribute to, for example, environmental sustainability depends entirely on how the terms are defined.[26]

In her book *Justice and the Politics of Difference*, Young argues that the meaning and shape of justice are contingent on the causes, shape, and consequences of *in*justices in the real world.[27] In contrast to political philosophers who sought to develop one universal, abstract theory of justice, Young contends that we must start by studying actually existing injustices in context, the social structures and institutions that uphold them, and the social movements that fight against them.

Young analyzes the claims and experiences of contemporary social movements pursuing social justice. Like many egalitarian political philosophers, she identifies the inherent injustice in material inequality and thus the need for more meaningfully distributive justice. Young and others also maintain that justice requires recognizing and redressing various forms of cultural oppression—the social relations and institutional processes that reproduce unequal distributive patterns over time and impede some people from standing as full members of society. Justice, as a result, requires recognition of the social structures that oppress certain social groups so that those groups can overcome the institutional subordination they experience.[28] Additionally, because unequal distribution and oppression fundamentally exclude certain social groups from full participation in politics, many scholars of EJ and other social movements argue that justice requires participatory parity. Finally, Amartya Sen, Martha Nussbaum, and other political philosophers have stressed that justice requires an adequate amount of capabilities—the basic institutions, resources, freedoms, and opportunities needed for people to be full members of society.[29] Key examples include jobs, living wages, clean air and water, and affordable and accessible public transit, health care, housing, and food.

Among the scholars who theorize EJ, many assert that it requires combinations of distribution, recognition, participation, and/or capabilities. For example, Cole, Foster, Hunold, Young, Kristin Shrader-Frechette, and Robert Lake emphasize distribution and participation; Figueroa points to distribution and recognition; and Pellow's work incorporates distribution, recognition, and participation.[30] In his book *Defining Environmental Justice*, Schlosberg contends that the EJ movement demonstrates a comprehensive notion of justice—one that joins distribution, recognition, participation, and capabilities—and that this is both a laudable and realistic way to address environmental injustices.[31] Schlosberg claims that the EJ movement itself illustrates how distribution, recognition, participation, and capabilities can be integrated—and that EJ activists also persuasively argue that effective environmental problem solving requires that they *must* be integrated.[32] He notes the "interplay" of these four components of justice—the ways in which they are mutually constitutive:

Not only are there different conceptions of justice apparent in the [EJ] movement, but the movement also recognizes that these notions of justice must be inter-related: one must have recognition in order to have real participation; one must have participation in order to get equity; further equity would make more participation possible, which would strengthen community functioning, and so on.[33]

In this book, I follow Schlosberg's lead in bringing together these four components of justice. Throughout the book, I use the concrete example of pesticide drift activism to illustrate the useful and necessary role that distribution, recognition, participation, and capabilities play in a socially just approach to environmental problem solving.

Yet the conceptions of justice that pesticide drift activists and the broader EJ movement work with as well as advocate are only half of the story. The EJ notion of justice must be understood not as appearing in a vacuum but in part as a response to the other conceptions of justice that shape and legitimize mainstream environmental politics (and politics in general), however incompletely, imperfectly, or unintentionally. Notably, although egalitarian ideals shaped the development of many crucial liberal political institutions in the Western world throughout the early and mid-twentieth century, the state's approach to addressing environmental problems has largely been forged by a utilitarian conception of justice.[34] Utilitarianism calls for maximizing welfare—where state interventions are part of providing the greatest good for the greatest number. Such a perspective justifies the widespread use of a cost-benefit analysis as the basis for environmental decision making in the United States today. David Harvey characterizes this as the "standard view" of environmental management. While recognizing the contributions to environmental protections that have been made under the standard view in the past century, Harvey points out that within a utilitarian framework, the "only serious question is how best to manage the environment for capital accumulation, economic efficiency, and growth."[35] In a context that privileges economic growth, the state is generally only able to intervene when there is quantified, certain scientific evidence documenting links between an environmental hazard and sufficiently egregious harm. Because this is essentially impossible for hazards whose impacts are realized unevenly across space and time, environmental problems have bloomed under the watch of utilitarian-based environmental regulatory apparatuses.[36]

In the wake of increasing critiques of the utilitarian, standard view of environmental management, two other conceptions of justice—libertarianism and communitarianism—have increasingly gained prominence over the past thirty years in mainstream Western environmental politics, and they both exacerbate problems like pesticide drift in important ways. Libertarians view individual liberty as the hallmark of justice, identify private property as the institution that best nurtures and protects liberty, and endorse the free market as the only socially just institutional mechanism of exchanging property. Communitarians argue that members of a "community" possess a shared understanding of the good life and thus are in the best position to identify their own conceptions of justice and injustice. Communities are said to reach these common understandings through tradition, shared experience, geographic proximity, and "relations of trust."

Throughout this book, I identify the increasingly significant roles that these theories of justice play in mainstream environmental politics, how those roles evolved, what the material consequences are, and how pesticide drift activists show that a radically different notion of justice is needed to effectively solve today's most pressing environmental problems. I show that libertarian and communitarian conceptions of justice increasingly influence and/or are reinforced by the efforts of the agricultural industry, the environmental regulatory state, and most agrifood activists to pursue environmental sustainability. I suggest that such ideas articulate with a context of considerable oppression and inequality in ways that reproduce—rather than alleviate—grave environmental problems like pesticide drift. I also highlight the mechanisms that make such ideas seem natural and reasonable, such as their propensity to displace environmental fallout to invisible bodies, to distant places, and into the future. As I will demonstrate, libertarian and communitarian theories of justice gain traction because the policies they inspire as well as justify allocate benefits largely to the relatively privileged, in turn deepening environmental inequalities at the same time that they claim to ameliorate them.

I maintain that the efforts by the agricultural industry, the environmental regulatory state, and alternative agrifood activists generally fail to adequately address the problem in part because they interact with libertarian and communitarian theories of justice. This is not to say that justice exclusively or even intentionally guides the efforts of industry, the state, or all activists, nor that these theories of justice are solely to blame for today's environmental problems. In fact, I spend considerable space in this book identifying the many other material and cultural structures that undergird predominant (and inadequate) approaches to solving environmental problems. My point is simply that identifying the theories of justice that justify and give traction to predominant solutions to environmental problems (and which, in turn, are reinforced by them) helps to explain the shape of mainstream environmental politics as well as its shortcomings—two critical and essential tasks in the broader move to more effectively and fairly solve present-day serious environmental problems. Throughout the book, I also showcase the work of pesticide drift activists, paying particular attention to the different vision of justice they advocate and the implications that poses for environmental illness and how we think about social justice. I find that effectively addressing environmental inequalities will require a state-society relationship that builds on the EJ notion of justice and strays wildly from the libertarian and communitarian ideas of justice that increasingly shape mainstream environmental politics today.

The Study

When I first began researching pesticide drift in 2001, I was eager to sift through the various proposed solutions and precisely identify the combination of policies and technologies that would solve the problem. After conducting my first round of interviews with people deeply invested in the issue, however, I discovered that the political conflicts over pesticide drift were fundamentally about the nature of the problem itself. Tremendous disagreement exists about every possible dimension of the problem, and ample evidence backs up each wildly different claim: how often pesticide drift occurs (rarely or daily?), how many people it affects each year (a few hundred or millions?), which people are most exposed (schoolchildren, farmworkers, or others?), what sorts of illnesses it causes (acute or chronic? minor or serious?), why it occurs (accident or inevitable?), and the state of scientific knowledge on which pesticide regulations are based (shoddy or robust?). Since crafting effective solutions requires that we first understand the nature of the problem, I thus shifted gears. I focused my attention instead on identifying the stories that people were telling about the problem itself, the evidence they used to back up their claims, and the context in which their ideas developed. Throughout my research, I found that stories revealed not only different ideas about pesticide drift but also fundamentally different notions of what justice looks like.

I concentrate on narratives—stories—to identify the points of contention in the public debates over pesticide drift. Narratives help to construct the world around us by defining what is possible and real. Environmental historian William Cronon emphasizes that narratives are a way to find values in a contradictory world—that we organize ecological change into stories with beginnings and endings in order to judge the morality of human actions.[37] Like other discourse analysts, Katherine Jones underscores that narratives have real, material consequences, as they help to shape the way that people understand the world around them:

It is the power of selection and simplification—or categorization—that gives representations their persuasive power. . . . They both encourage certain meanings and constrain or limit other meanings. . . . [T]he rules of social order and the practices of representation go hand in hand.[38]

Michel Foucault, whose research dramatically challenged the way we understand the relationships between power and discourse, acknowledges that while some discourses reinforce the status quo, others explicitly contest inequalities:

We must make allowance for the complex and unstable process whereby discourse can be both an instrument and an effect of power, but also a hindrance, a stumbling block, a point of resistance and a starting point for an opposing strategy. Discourse transmits and produces power; it reinforces it, but also undermines and exposes it, renders it fragile and makes it possible to thwart it.[39]

The purpose of my project, as with much other academic work, is to search for "other," subordinated narratives along with the suggestions they present for the making of a more socially just and ecologically sustainable society. I showcase direct statements from the actors themselves because they illustrate the hegemonic and marginalized narratives about pesticide drift. These direct statements exemplify common arguments and claims. Accordingly, they should be understood as representative of broader patterns, and unique only in terms of their clarity and brevity. In presenting the two main stories told about pesticide drift, my goal is not to determine which one is correct. Rather, I believe that they are both valid but incomplete. Each one offers important insights and its own partial perspective into a complicated, imperfectly knowable world. Each story highlights certain bits of information and sidelines others, privileging some issues over others.

Throughout the book, I emphasize two conclusions that I draw from this narrative analysis. First, the predominant story told about pesticide drift does not account for the ways in which social inequalities and oppressive social relations contribute to pollution and illness; it in fact ignores those relationships and renders them invisible. I argue that effective environmental problem solving must meaningfully account for the effects of inequalities and oppression on our abilities to understand as well as prevent problems like pesticide drift along with the grave illness and fear they add to. Second, I argue that the conflict between the two narratives serves as a window into competing conceptions of social justice. The debates over pesticide drift provide us an opportunity to critically reflect on the strengths and weaknesses of the particular visions of justice that govern politics today, not to mention those that could possible help us address environmental and other problems more effectively.

To determine the stories that people tell about pesticide drift, from 2001 to 2009 I gathered data from multiple sources: interviews, observation, and printed materials. I draw heavily on over a hundred in-depth, loosely structured interviews that I conducted with regulatory officials, research scientists, pesticide drift activists, and other agricultural community residents. In analyzing those interviews, I sought to determine how various actors define pesticide drift, how they frame the scope and scale of the problem, what an appropriate set of solutions would be, and what sources of information they draw on to make their decisions and claims. I also asked these questions while examining published materials from regulatory agencies and activist groups, reading newspaper reports of pesticide drift events, and observing key events like activist demonstrations and regulatory hearings. To situate these various stories within their political-economic context, I also use secondary data on pesticide use patterns and demographic change, and also historical accounts of California agriculture, farm labor, pesticide activism, and regulatory reform in California.

Throughout the book, all unreferenced quotes are excerpts from my own interviews. In this book and my other published work, I typically obscure the identity of the individual speakers. I do this for two reasons: to protect a subset of my informants for whom speaking candidly with me could put them in jeopardy, and to focus the reader's attention on the narratives and institutional practices rather than the individual people.

This case study focuses on the southern end of California's Central Valley. I targeted this region for several reasons: large-scale pesticide drift incidents have occurred there on a regular basis more than in any other part of California; use rates of the most toxic pesticides there consistently rank among the highest in the state (and the nation); its air pollution ranks among the worst in the nation, rivaling that of nearby Los Angeles; people who suspect they have been exposed to pesticides consistently report being ignored by regulatory officials; and the region is in a state of "significant economic distress," with the average per capita income well below state and national averages.[40] Also, the Central Valley receives comparatively little attention from academic researchers, the state, or the general public; it is in many ways the "forgotten" California. The Central Valley is neither beaches nor mountains, but the space in between; interstate traffic races across or gingerly along the edge of this landscape to reach more scenic and cosmopolitan destinations. My motivation to focus on the Central Valley was fueled by residents' repeated assertions that this place has become, as one said, "California's dumping ground"—similar to the "sacrifice zones" elsewhere that Robert Bullard, Valerie Kuletz, and other scholars have studied.[41] Despite being home to the state's highest population growth rates along with a number of new bedroom communities to Los Angeles and the San Francisco Bay Area, the Central Valley appears to many residents as having been abandoned as a wasteland of toxic freeways, agriculture, waste incineration, and megadairies. I hope that this book will help to bring constructive yet critical attention to this and other zones of sacrifice in ways that honor the diversity, dreams, skills, and rights of its residents as well as the ecologically sustainable spaces that these landscapes can become.

The choices I have made here—which case to study, how to analyze it, and how to write about it—are all made in the spirit of critically and normatively evaluating the status quo, like most of the scholars I reference throughout. Young phrased this motivation nicely in her introduction to *Justice and the Politics of Difference*:

Social description and explanation must be critical, that is, aim to evaluate the given in normative terms. Without such a critical stance, many questions about what occurs in a society and why, who benefits and who is harmed, will not be asked, and social theory is liable to reaffirm and reify the given social reality.[42]

The Book's Organization

In chapter 2, I make my case that pesticide drift is a dangerous socio-environmental problem worthy of interrogation. I draw on a wide range of social and technical data to explain why pesticide drift is best understood as a case of widespread yet generally invisible and ignored chemical contamination. From there, I turn to analyzing the three social groups that most directly shape the problem of pesticide drift: the pesticide industry, the environmental regulatory state, and alternative agrifood activists. In chapters 3, 4, and 5, I describe those actors' efforts to address the problem of pesticide drift and why those efforts have generally failed, identifying in particular the specific ways in which libertarian and communitarian theories of justice undergird many of those efforts and undermine their effectiveness.

I start with industry, in chapter 3, to explain how chemical pesticides became the predominant model of agricultural pest management—how a diffuse network of industry actors, each with varying degrees of political-economic power, invested deeply in the "pesticide paradigm" of agricultural pest management. I then describe industry's efforts to address pesticide drift, look at the industry actors' shortcomings, identify their particular intersections with a libertarian notion of justice, and explore the work of pesticide drift activists to confront industry's culpability in the problem of pesticide drift.

In chapter 4, I turn to the attempts of environmental regulatory agencies to step in and control pesticide drift in ways that industry cannot. I draw critical attention to how libertarian and communitarian theories of justice shape trends in environmental regulation and effectively undermine regulatory efforts to control the problem of pesticide drift. I conclude that chapter by showcasing the ways that pesticide drift activists confront pesticide regulatory agencies, and I indicate how their demands and practices implicitly criticize libertarian and communitarian theories of justice while simultaneously articulating an entirely different notion of what justice means. Here, I also detail pesticide drift activists' policy prescriptions and introduce the precautionary principle as an overarching framework for putting pesticide drift activists' theory of justice into practice.

In chapter 5, I then examine the alternative agrifood movement's efforts to critique and ameliorate the failures of both industry and the state. Pesticide drift activism overlaps with other branches of alternative agrifood activism in terms of individuals and institutions. However, I emphasize the distinction between pesticide drift activism and the predominant branches of the alternative agrifood movement in order to highlight some notable differences between the two groups (especially in terms of the contrasting models of change that they prioritize and the different theories of justice that guide their work). I explain why the alternative agrifood movement's priorities and practices have historically sidelined the problem of pesticide drift experienced in agricultural communities, focusing on these activists' material constraints as well as the ideological adherence by some key elements of the alternative agrifood movement to libertarian and communitarian theories of justice. I conclude that chapter by describing the ways that pesticide drift activists manage strategic relationships with the sustainable agriculture movement as well as other social movements, demonstrating how their approach to activism contains a critique of the theories of justice underlying most of the alternative agrifood movement and an entirely different notion of what justice means.

Lastly, in chapter 6, I summarize the factors that explain why the problem of pesticide drift festers despite various actors' environmental sustainability efforts, evaluate the achievements of pesticide drift activism (both in terms of material accomplishments and its role in casting EJ in a new light), and make several specific policy recommendations based on what the case study tells us about what justice requires. In particular, I focus on the notion of "institutionalizing" EJ—integrating the EJ movement's theory of justice into the everyday work of environmental regulatory institutions. I examine in more detail the precautionary principle as a framework for doing exactly that.

2

Assessing the Scope and Severity of Pesticide Drift

In this chapter, I illustrate why I find pesticide drift to be an egregious problem worthy of critical investigation. To do this, I showcase a wide array of technical evidence of pesticide illness and drift. Understanding the scope of pesticide drift requires that we evaluate the various strengths and weaknesses of those technical data explicitly in terms of the broader social context in which they were collected and by which they are profoundly shaped. I therefore start the chapter with an overview description of California agriculture—paying particular attention to the technological, cultural, and social structures that give rise to its industrial character. I then turn to the task of assessing the frequency, scope, and severity of pesticide drift. I examine many different types of data from a range of sources and disciplines and critically evaluate them in light of the social and cultural contexts within which they are embedded. As will become clear, the widespread assertion that pesticide drift is a minor and controllable issue is simply untenable. This chapter thus sets the stage for the rest of the book, in which I explain how the problem of pesticide drift came to be both pervasive and invisible as well as how a small group of activists is working to change that.

California Agriculture: A Technological, Cultural, and Social Project

At first glance, California agriculture is a portrait of abundance, as it provides half of the United States' fruits and vegetables and generates $38 billion in annual sales. High-value fruit, nut, and vegetable crops have come to predominate agricultural production in California—a direct artifact of the region's mild climate, the state's massive irrigation network, the public university research and extension service priorities, and private investors' forays into chemical and mechanical farm inputs and transportation, processing, and marketing networks.[1] The production of high-value crops has become capitalized in the form of high land values—a process that has been simultaneously reinforced in much of the state by suburban development and other population growth pressures on agricultural land values.[2] Consequently, growing low-value crops is simply not economically viable throughout most of California.

In addition to rising land values, increasing costs of production (e.g., seeds, machinery, and insurance), competition by foreign growers, and increasing concentration in downstream sectors (such as processing and retail) put downward pressure on farmers' profits.[3] The predominant strategy by which growers have sought to maintain profits in such a context is by intensifying production—employing chemical, mechanical, and labor inputs to increase the productivity per acre to the greatest extent possible.[4] California fields exemplify this industrial approach to agricultural production, and this is strikingly evident in terms of pesticide use. California agriculture accounts for only 2 to 3 percent of all U.S. farmland, but 25 percent of the nation's agricultural pesticides.[5] This form of productive, high-value agriculture has been made possible by extraordinary environmental exploitation. Intensifying production through using high rates of fertilizers and toxic pesticides creates many forms of water and air pollution, which are paid for by humans and other species through illness and habitat destruction. Moreover, in this economic context, growers have little incentive to regularly rotate in less profitable cash crops or cover crops that would help to regenerate the soil, reduce pest pressures, decrease chemical use, and provide other ecological services. High rates of pesticide use in such overly simplified farming systems also create a "pesticide treadmill," where pest populations adapt, proliferate, and compel the development of new chemical solutions.

It is worth recognizing that all agricultural fields are highly simplified ecological systems: in growing one (or a few) plants, all other plants and many other forms of life (insects, birds, and other animals) become "pests"—a socially constructed category of living organisms that threaten agricultural productivity and profits. Many pests thrive in agricultural environments, and so pest control of some form or another is an integral component of any agricultural system. In chapter 3, I explain how highly toxic chemical pesticides have come to form the foundation of pest control throughout the world today. Though commonly referred to as "conventional" agriculture, chemical-intensive agriculture is a distinctly "modern" phenomenon, and is best understood as the outcome of specific historical contingencies rather than inevitable or inherently rational. Many "alternative" approaches to pest management avoid chemical pesticides (or use them only as a last resort), honor ecological principles and relation-ships, have deep historical roots, and continue to form the basis of many highly productive and innovative agricultural systems around the world today.[6] One of the purposes of this book is to illustrate the consequences of the predominance of the chemical-intensive model of agricultural pest management.

The industrial model of agriculture so earnestly pursued in Califor-nia—the extraordinarily simplified and engineered ecosystems of mono-cropped fields, intensified production, rerouted waterways, and chemical controls—has been as much a cultural project as a material one. Historian Christopher Bosso uses the term "pesticide paradigm" to characterize the "unspoken and almost unconscious assumptions" by industry, the state, and the public alike that chemical pesticides are absolute necessary, safe, and effective—a set of beliefs that shaped the notable degree to which pesticides have been embraced, defended, and minimally regulated. As Bosso explains, in the early part of the twentieth century, pesticides were "equated predominantly with agricultural progress, while most other concerns, such as 'the environment' (a term not yet in use) or 'safety,' did not intrude so mightily into the debate."[7] The pesticide paradigm is a central component of what DuPuis calls the "enlightenment" model of modern industrial agriculture: the investment of extraordinary resources into technologies and knowledge developed by formally trained, "ratio-nal" experts in order to grow specialty crops on an industrial scale. As DuPuis describes it, the enlightenment model of industrial agriculture has been buoyed by a decidedly modernist vision of progress. In this vision, the modernizers (including economists, farm advisers, bankers, and ag-ricultural input suppliers) have championed technological developments as essential to controlling the risks that nature posed to agricultural pro-duction; this productivity, it has been believed, yields not simply highly productive farms but also social progress and freedom from deprivation.[8] Conflicts over pesticide drift show the endurance but also the fallacy and consequences of this belief, illuminating the uncontrollable environmental life of chemical pesticides and the ecological price on which industrial productivity has blossomed.

The productivity of California agriculture has been predicated not only on ecological exploitation but on social exploitation as well—notably, of immigrant farmworkers who have been recruited to work in California's fields for 150 years. Agricultural production in California and increasingly throughout the rest of the United States pivots around a ready supply of cheap, flexible workers willing to work when needed and disappear when the (often-temporary) job is done. As a result, approximately seven hundred thousand hired farmworkers populate California's agricultural landscape.[9] Social science survey work has found that over 90 percent of California farmworkers today are Mexican immigrants, most live in poverty (with over half of all farmworking families earning less than fifteen thousand dollars per year, due largely to the seasonal and thus inconsistent nature of agricultural jobs as well as low wages), few make use of any needs-based social services, and at least half lack legal authorization to be in the United States.[10]

To a large extent, immigrant farmworkers' economic and political vulnerabilities have been actively constructed. The state and the agricultural industry both played important roles in recruiting various immigrant labor groups to work on California farms and oppressing them in the process.[11] Up through the 1960s, this was accomplished through the direct, systematic overrecruitment of different ethnic-based migrant labor groups, curtailing those workers' legal rights and compelling them (through physical force or otherwise) to disappear when the seasonal work was completed. These strategies effectively pushed down agricultural wages, established elaborate migration networks (where those between Mexico and the United States predominate today), and shifted the costs of the reproduction of labor to other places.[12]

Social scientists have discussed how the militarization of the U.S. border since the mid-1990s (which escalated to an unprecedented degree during the presidential administration of George W. Bush, with its post-9/11 flurry of institution building conducted in the name of the "war on terror") further increases immigrant vulnerability.[13] The militarization of the U.S.-Mexico border includes the massive increases in U.S. Border Patrol personnel (from around three thousand agents in 1985 to over twenty thousand in 2010), expansion of various border walls (now covering 520 miles of the border), expanded investment in traditional border technologies (trucks, helicopters, and night-vision equipment), as well as a new generation of surveillance technologies (unmanned aircraft and underground sensors) that cover the landscape.[14] Researchers have shown that the increasingly militarized border does not decrease the factors that drive migration but simply shifts migration to less visible spaces (notably, remote mountains and deserts) and catalyzes the development of a network of coyotes (human smugglers). Migration across the U.S.-Mexico border has therefore become more expensive and dangerous, as evidenced by the hundreds of migrants who tragically die in the deserts and mountains of the U.S.-Mexico border zone every year, and the estimate that such deaths doubled between 1995 and 2005.[15]

In the past few years, immigration policy enforcement has extended beyond the border zone itself, and now includes increased federal enforcement in the interior of the United States, the devolution of sanctioned enforcement to local law enforcement agencies, and the rise of informal/ private enforcement by vigilante groups like the Minutemen.[16] These expanding enforcement practices throughout the interior underscore the value of conceptualizing the U.S.-Mexico border as "floating" and having a pervasive presence in immigrants' lives. It is for this reason that anthropologist Lynn Stephen has said that unauthorized immigrants in the United States live under a "gaze of surveillance."[17] Such increasingly pervasive immigration enforcement constrains the ability of unauthorized immigrants—and people with unauthorized friends and family—to seek legal protections or use basic social services. This deepening marginalization in turn thwarts the ability of social service providers to track, understand, and treat issues in and beyond the workplace, such as health and safety concerns, physical abuse, economic exploitation, and other crime.

In addition to these institutions that deepen the inequalities that immigrant communities experience, many cultural phenomena render those injustices invisible or unremarkable. In some cases, blatant, racialized narratives of difference (e.g., "Oaxacans like to work bent over") legitimize exploitative agricultural labor relations.[18] Researchers have also shown how seemingly innocuous ideas and narratives function in similarly oppressive ways. For example, Raymond Williams and Don Mitchell have long shown how popular conceptions of rural landscapes depoliticize agricultural labor and render it invisible.[19] Leo Chavez has demonstrated that Americans tend to conceptualize the relationship between the United States and Mexico as a flow of problems only going from south to north (e.g., immigrants represent a "flood" or "invasion," immigration is a "crisis," and the United States appears neutral and passive), with little recognition of the culpability of the United States in Mexico's underdevelopment, or the wide range of positive contributions that Mexicans make to the U.S. economy and society.[20]

Although immigrant farmworkers and their family members are certainly not the only California residents exposed to pesticide drift, they are disproportionately exposed to agricultural pesticides at work and by living near agricultural fields. Pesticide drift incidents simply cannot be understood apart from the relations of oppression that characterize immigrant farmworking communities. In interviews with me, residents of the Central Valley towns in which major pesticide drift incidents have occurred report that the majority of the households contain farmworkers. U.S. Census Bureau data indicate that these communities are on average poor and populated predominantly by Latinos.[21] Unemployment rates in the southern counties of California's Central Valley ranged from 15 to 20 percent in 2010—well above the national average.[22]

The social issues facing farmworkers and other disadvantaged residents make these people not just disproportionately exposed but also disproportionately unable to deal with actual pesticide exposures, since many of the conditions they experience limit their ability to report exposure, get medical treatment, and/or pursue legal claims. Many farmworkers avoid reporting pesticide exposure for fear of retribution from employers, particularly in light of the general overabundance of people seeking farmwork.[23] For families lacking health insurance and living on a low income, medical and legal services are also simply unaffordable. Additionally, when people who suspect that they have been exposed to pesticides are legally unauthorized—or who live with unauthorized relatives or friends—they are often unwilling to report pesticide exposure or access any social services for fear of interacting with law enforcement and thereby risking apprehension. Social issues like poverty, language barriers, and legal status consequently deepen the problem of pesticide drift and also make it more difficult to accurately quantify.

The social inequalities facing immigrant farmworkers and other residents of agricultural communities must be understood not only in absolute terms but in relative ones as well. Relative to agricultural and chemical industry actors, any individual exposed to agricultural pesticide drift and/ or concerned about the issue has little power to shape public understanding, decision-making processes, or the environment in which they live. It is within this context of considerably asymmetrical power relations that agricultural pesticide use, drift, illness, regulation, and political conflict unfold. Understanding these various social relations will help us evaluate the technical evidence of pesticide use, drift, exposure, and illness.

Agricultural Pesticide Use in California

Because pesticide drift is, to a considerable degree, a function of pesticide use, data on pesticide use rates and application methods provide one general indication of the scope of pesticide drift. California's unique Pesticide Use Reporting (PUR) system data detail the types and quantities of pesticides used within the state and the methods by which they are applied, which give some indication about their drift potential and toxicity. Approximately 90 percent of all pesticides used in California are prone to drift, due to the application method (e.g., aerial spraying) or product formulations (e.g., dusts or high volatility).[24] They are thus likely to move far from the site of application, which occurs during the time of application or over the course of several days. This drift happens in many ways; a liquid spray being carried by the wind ("spray drift") is a well-documented and relatively visible phenomenon.[25] Sprays also can attach to soil particles, which are then carried by the wind. Soil fumigants (notably, methyl bromide, metam sodium, metam potassium, chloropicrin, and 1,3-dichloropropene) are among the chemicals that are designed to turn into a vapor in order to permeate and sterilize the soil structure; these vapors easily move away from the site of application.[26]

In addition to using large quantities of drift-prone pesticides, California agriculture relies extensively on pesticides that are particularly toxic to humans. It is worth emphasizing that pesticides vary tremendously in their human toxicity. Notably, pesticides vary in terms of their acute toxicity (ability to cause short-term illness), which can range from mild to severe. Many pesticides also contribute to chronic toxicity (causing long-lasting or frequently recurring illness). Major chronic illness categories include cancer, reproductive and developmental problems, endocrine disruption, respiratory disorders, and immune system depression.

Different pesticides used in California agriculture thus pose distinct threats to public health. Some widely used pesticides are prone to drift but are only mildly acutely toxic; sulfur dust is a classic example of such a pesticide. In this book, I focus instead on chemicals that are used in high quantities, are prone to drift, and are highly toxic (in terms of acute and/ or chronic illness). The use of highly toxic and drift-prone pesticides is especially concentrated in the southern end of California's Central Valley, as indicated in figure 2.1 below. The region's particular geographic and atmospheric conditions (e.g., basin shape, inversion layer, and heat) trap pesticides and other air pollutants and make them volatilize quickly, thereby exacerbating the propensity for pesticide drift events and exposures in this region.

I pay particular attention to soil fumigants throughout this book, and my examples will often pertain to fumigants. Chemical companies began widely promoting soil fumigants in the 1940s for controlling nematodes, plant pathogens, and weeds. Soil fumigation has become the standard preplanting treatment in many commodity sectors plagued by these pests (notably, strawberries, tomatoes, tobacco, and root crops like carrots,

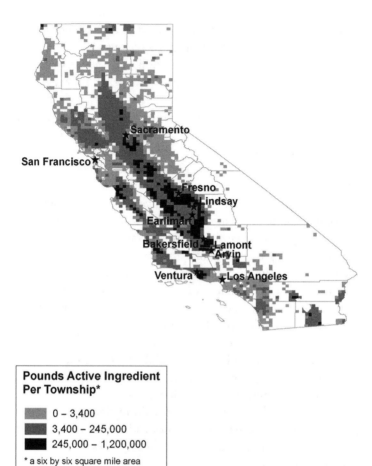

Pounds Active Ingredient Per Township*

- 0 – 3,400
- 3,400 – 245,000
- 245,000 – 1,200,000

* a six by six square mile area

Figure 2.1

Map of reported "bad actor" pesticide use intensity in California in 2008, by township. *Note:* This map reflects the total number of pounds of active pesticide ingredients applied per township (a six-by-six square mile area); the darkest areas therefore reflect the spaces of highest pesticide use. The areas of light gray indicate the townships with 1–3,400 pounds (0th–50th percentile); the areas of dark gray indicate townships with 3,400–245,000 pounds (50th–90th percentile), and the black squares show those townships with 245,000 or more pounds (90th–100th percentile). *Sources*: California DPR's PUR data from 2007; map created by Joshua Pepper, Pesticide Research Institute, Inc.

peanuts, and potatoes), replacing crop rotations and other less toxic, ecological approaches to pest management.[27] Fumigant use in the United States is particularly concentrated in California (as shown in figure 2.2 below), largely because of the crops grown there.

With the exception of horticultural oil and sulfur (which pose relatively low risk to human health), fumigants are used in greater absolute quantities than any other group of pesticides in California and account for 20 percent of all agricultural pesticides used in the state.[28] Although the use of each fumigant has shifted slightly over time, total soil fumigant use has not declined (see figure 2.3 below).

Designed to vaporize so they can permeate and sterilize soil, fumigants are also highly prone to drift. Their application methods can exacerbate their drift potential and are determined by local, state, and federal

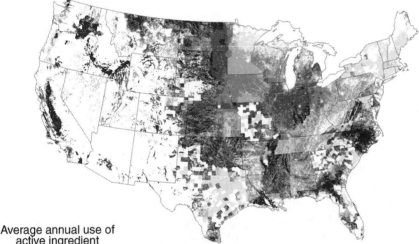

Average annual use of
active ingredient
(pounds per square mile)

- no estimated use
- > 0 - 0.46
- > 0.46 - 2.4
- > 2.4 - 6.1
- > 6.1 - 13
- > 13

Crops	Total pounds applied	Percent national use
potatoes	42293883	37.67
tomatoes	12553325	11.18
tobacco	11737029	10.45
strawberries	8741320	7.78
carrots	6197561	5.52
peanuts	6171386	5.50
bell peppers	4111515	3.66
dry onions	3832686	3.41
sugarbeets	3092641	2.75
cucumbers and pickles	1958654	1.74

Figure 2.2
Estimated annual agricultural use of soil fumigants for 2002. *Sources:* Preparation of data and graphical analysis by U.S. Geological Survey using methods described in Thelin and Gianessi (2000). See also U.S. Geological Survey (2011).

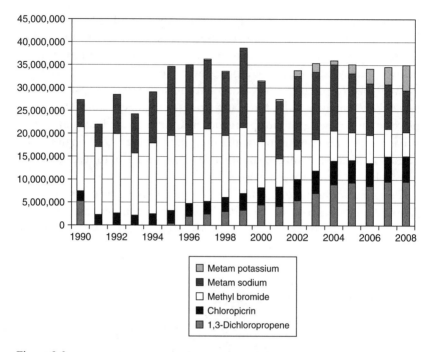

Figure 2.3
Agricultural use of fumigants in California, 1990–2008, measured in pounds of active ingredient. *Note:* Fumigant use here refers to total pounds of active ingredients reported for agricultural purposes through the California DPR's PUR system.

regulations. In many cases, fumigants are applied via sprinklers—a method that drastically exacerbates the drift of volatile pesticides like fumigants. In other cases, fumigants must be applied via "shank injection" (where a tractor mounted with special equipment injects the fumigant under the soil surface) or drip irrigation; typically these methods are followed by "sealing in" the fumigant with sprinkler irrigation or plastic tarps. In addition to their high acute toxicities, fumigants and/or their breakdown products pose many serious chronic health threats to humans. For example, 1,3-dichloropropene is firmly recognized as a carcinogen; chloropicrin causes chronic respiratory damage and is a carcinogen; metam sodium and metam potassium are carcinogens and developmental toxicants; and methyl bromide is a neurotoxicant and developmental toxicant.[29]

The other major group of pesticides featured in this book is organophosphates, especially the widely used chlorpyrifos and diazinon because of their high use rates, high acute toxicity in humans, association with chronic illness, and frequent cause of illness in pesticide drift events.

German scientists initially developed organophosphates in the early twen-tieth century as nerve gases for chemical warfare, and in the postwar years, U.S. corporations adapted and marketed the compounds for agricultural purposes. Organophosphates function in the human body much as they do in insects: as neurotoxicants, they disrupt the normal transmission of nerve impulses, and in sufficient quantities, lead to paralysis and death. Widely used because of their effectiveness and low environmental persis-tence, many organophosphates and their breakdown products are highly acutely toxic and implicated in chronic illness as well, even at low levels of exposure (notably, neurological degeneration and impairment of brain development in fetuses, infants, and children).

Evidence of Pesticide Drift

My concern in this book is not only with the use of agricultural chemicals per se but also with their movement away from the site of application and their subsequent contribution to human illness and other environmental problems. The fact that pesticides move off-site is in fact not in dispute: it is widely acknowledged by regulators, industry, and activists alike that pes-ticides will drift away from their site of application to some extent.[30] The task at hand is to evaluate the extent to which pesticide drift constitutes a threat to human health. Doing so requires that we determine whether drifting pesticides exceed a level that people can safely be exposed to and what the consequences are of those trends. As I discuss in this section, various sources of data indicate that pesticide drift widely occurs at levels that threaten human health.

Illness Data

Illnesses experienced in association with specific pesticide applications are the most direct and dramatic indication that pesticides drifted off-site at problematic levels. California regulatory agencies have tracked pesticide illness reports since the early 1970s.[31] Today, California DPR's Pesticide Illness Surveillance Program (PISP) tracks all reported pesticide exposures in the state and classifies those illness cases according to their degree of supporting evidence, the likelihood that they are a result of a pesticide application, the type of application, and many other variables. In the most recent ten-year period for which data are available (1998–2007), the PISP system documented an average of 261 illnesses annually as be-ing possibly, probably, or definitely caused by agricultural pesticide drift (see figure 2.4).[32] The counties that lead the state in terms of pesticide

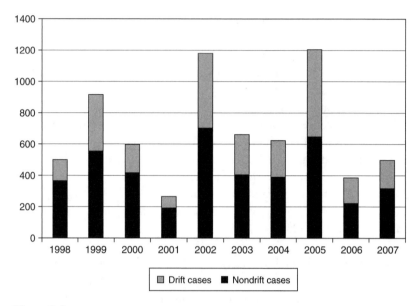

Figure 2.4
Annual number of reported and confirmed agricultural pesticide illnesses, 1998–2007. *Note:*
The chart shows the number of reported and confirmed individuals (cases) made ill by exposure to agricultural pesticides. *Source:* California DPR's PISP system.

use also accounted for the majority of the pesticide illnesses over the past
ten years (see table 2.1). PISP data associate 55 percent of agricultural
pesticide drift illness cases with fumigants and another 20 percent with
major organophosphates.[33]

Outside of California, several U.S. states have pesticide illness tracking
systems, supported at least in part by the U.S. Environmental Protection
Agency (EPA).[34] They have been characterized as "rudimentary and/or
poorly enforced" relative to California's, notwithstanding the critiques
of the latter.[35] The U.S. EPA maintains no comprehensive database on the
national extent of reported pesticide illnesses from pesticide drift or any
other source, and its efforts to cobble together various inadequate data
sets to estimate national acute pesticide illnesses have been critiqued as
flawed on numerous grounds.[36]

For many reasons, reported pesticide illness data actually greatly
underrepresent the extent of pesticide exposures and associated serious
illnesses. Exposed individuals often do not report acute pesticide illness because they do not attribute their symptoms to pesticide exposure
(since acute symptoms of pesticide exposure frequently include nausea,

Table 2.1
Total reported and confirmed illness cases from agricultural pesticide drift from 1998–2007 in California's highest pesticide-using counties

County	Number of illness cases	Rank in pesticide use
Kern	835	2
Monterey	598	6
Tulare	337	3
Fresno	109	1
Kings	69	10
Ventura	67	8
Merced	58	7
San Joaquin	57	4
Madera	53	5
All other 42 counties	*384*	

Sources: California DPR's PISP system; California DPR's PUR summary data from 2007.

headaches, burning eyes, and other common ailments), do not know how to report their illness to authorities, or do not experience any notable acute symptoms of exposure (since many of the health effects are only manifest over the long term).[37] Additionally, most health care providers have little training in recognizing, treating, and reporting pesticide exposures, and the process of reporting pesticide exposure involves navigating a complicated bureaucratic maze.[38] Moreover, as I elaborate in chapter 4, many people who suspect they have been exposed to pesticides are afraid to report or lack the means needed to substantiate their claims.

For pesticide exposure cases that are reported, the available evidence determines whether the illness is classified as definitely, probably, possibly, indirectly, unlikely, or not related to pesticide exposure.[39] Qualifying an illness as definitely resulting from pesticide exposure requires both medical evidence (characteristic signs observed by a medical professional) and physical evidence of exposure (such as samples of contaminated clothing). For this to happen, the exposed individual must go to a health clinic quickly and insist that the clinic conduct the appropriate medical test of pesticide exposure, such a test must exist, that facility must have the resources to conduct the test, the doctor must contact the public health department, the public health department must contact the agriculture commissioner, and agriculture commissioner staff members must respond quickly to collect soil or foliage samples that could confirm drift of the pesticide in question before the product breaks down or moves further off-site. PISP data illustrate the difficulty of being able to definitively determine an illness as pesticide related: only 5 percent of agricultural pesticide illness cases from 1998 to 2007 are classified as definite. I include statistics about probable and possible cases, since they are defined as those where illness and exposure information are strongly correlated to, but not fully substantiated by, incontrovertible evidence.

These very factors that prevent most reported pesticide illness cases from being classified as definite also prevent most known pesticide illness cases from being reported at all. As a result, reported pesticide illness data greatly underestimate the extent of acute illness from pesticide drift. Interviews with farmworkers and other agricultural community residents demonstrate that a sizable percentage of these people have been exposed to agricultural pesticide drift. For example, in a recent survey of 321 residents of one farmworking community conducted by CPR, 41 percent of the respondents reported that they had been "drifted on," and more than half of those individuals stated that this had happened to them two to five times.[40] As one farmworker advocate stated, "If you ask a roomful of farmworkers how many of them have been involved in a pesticide drift, at least half will raise their hands."[41] Perhaps most important, pesticide illness data are not a good proxy for the actual incidence of drift since they only reflect some acute illnesses—but not any chronic illness associated with pesticide exposure. Indeed, exposure to many pesticides can lead to chronic health problems, but may not produce any acute symptoms at all. Reported pesticide illness data should thus be interpreted as the tip of the iceberg of actual pesticide illness.

Monitoring Air, Water, and Other Media

In contrast to pesticide illness data, air-monitoring studies are a controlled and direct method of testing for as well as documenting pesticide drift. In such studies, researchers set up specialized air-monitoring devices that include a vacuum pump that pulls air at a constant rate through glass sampling tubes, which are filled with a special pesticide-trapping resin or other filter. Researchers later analyze the concentration of one or more specific pesticides trapped in the filter, and then use the contextual data (e.g., the flow rate of the pump and the length of the time sampled) to calculate the overall ambient concentration of those pesticides during each sampling period at the study site. Scientists compare air-monitoring data to some sort of health benchmark (a particular amount of pesticides in the air, below which no adverse health effects would be expected) in order to determine whether current pesticide use practices pose undue risks to public health.

Air-monitoring study results vary widely and are shaped by many variables, including the timing of the study, the current pesticide use patterns, the location of the monitoring devices, and the researchers' adherence to sampling protocol. California's DPR and Air Resources Board (ARB) together conduct the majority of the pesticide air-monitoring studies in the state. Each year, DPR and ARB conduct "ambient" and/or "application site" types of air monitoring for several agricultural pesticides. In the ambient monitoring studies, monitoring devices are placed at several locations in agricultural communities (typically atop schools or other public buildings for about four weeks) during a season of anticipated peak use of the target pesticide to assess the general population's exposure. In the application site studies, air samples are taken at several edges of a field for up to seventy-two hours during and after a specific pesticide application to assess the maximum short-term concentrations to which the public could be exposed.

DPR's summaries of its own air-monitoring studies present the airborne concentration results. Publicly available DPR documents interpret those results only for those pesticides undergoing special review as suspected Toxic Air Contaminants (TACs), and those analyses have determined that numerous widely used pesticides (including the fumigants chloropicrin and metam sodium) exceed DPR's health benchmarks.[42] Researchers at PAN analyzed many of the other ARB/DPR air-monitoring results (i.e., for pesticides that have not gone through TAC review), and found that the airborne concentrations of several major soil fumigants (including methyl bromide and 1,3-dichloropropene) and other selected pesticides (including chlorpyrifos, diazinon, and chlorothalonil) exceeded various health benchmarks, frequently by large margins.[43] Similarly, activist researchers in Washington State recently analyzed the results of ambient and application site air monitoring done there by the state's Department of Health from 2005 to 2007 and concluded that airborne concentrations of methyl isothiocyanate (MITC), a toxic breakdown product of two major soil fumigants, routinely exceeded various health benchmarks.[44]

Researchers outside regulatory agencies have conducted their own air-monitoring studies and have found problematic airborne concentrations. Pesticide Research, Inc., conducted air monitoring in Moss Landing, California, near a fumigant application and discovered levels of chloropicrin that exceeded health benchmarks.[45] PAN's Drift Catcher air-monitoring program is functionally equivalent to the DPR/ARB system, and the Drift

Catcher studies have found ambient concentrations that exceeded health benchmarks in their monitoring studies of chloropicrin in Sisquoc, California; chlorpyrifos in Washington State; and chlorpyrifos in Lindsay, California.[46] Moreover, urine testing conducted by PAN showed that a sample of the individuals living at the location of the chlorpyrifos air monitoring in California had problematically high concentrations of the pesticide's metabolite in their bodies.[47]

Additional research sketches a broad picture of pesticide drift as pervasive and long-distance. Organochlorine pesticides and other environmentally persistent chemicals volatilize from where they are applied, travel through atmospheric air currents, and are deposited in the colder regions of the globe—thousands of miles from where they are manufactured and used.[48] In California, agricultural pesticides have been detected in rainwater and fog and are routinely found at levels that exceed regulatory thresholds in surface and groundwater.[49] In one watershed where pesticide concentrations in the surface water exceeded the regulatory limits, the U.S. Geological Survey attributed the majority of the pesticide loading to the volatilization of pesticides from nearby agricultural applications.[50] Other researchers found agricultural pesticides in the Sierra Nevada mountains, up to fifty miles from the nearest agricultural fields.[51]

Monitoring studies thus show that pesticides drift long distances and often at concentrations that exceed levels of health concern. Yet they only show that pesticide drift actually occurs at problematic levels—not how frequently or regularly this happens. Ambient concentrations of pesticide drift into agricultural communities are simply not constantly (or even regularly) monitored. In fact, DPR has conducted ambient air-monitoring studies for an average of only five pesticides per year and application site monitoring studies for an average of only three pesticides annually since 1986—out of nearly one thousand pesticide active ingredients registered for use.[52] Overall, since 1986, DPR and ARB have conducted ambient monitoring for sixty-eight different pesticides, and application site monitoring for forty-five different pesticides; typically each of those pesticides has been monitored once, although some problematic pesticides have been monitored multiple times (notably, the fumigants in recent years). The limited amount of pesticide monitoring stems to some extent from its resource-intensive nature: laboratory analysis and sampling equipment can be costly, only a limited number of pesticides can be tested for in each sample, and the work requires considerable labor from trained staff. In short, there is no equivalent of a litmus test or other magic wand for monitoring airborne pesticides.

As a consequence, however, the risk of exposure to pesticide drift is poorly understood.

Air-monitoring studies all have additional limitations. Studies at the edge of application sites are conducted with the permission of and in cooperation with the landowner and pesticide applicator, thus obviously representing applicators' most careful practices, and not necessarily typical behavior or practices for that pesticide or application method. Also, linking the timing and physical placement of the ambient studies to the target pesticide's location and season of peak use can be tricky. Low ambient levels in the study may as a result represent poor timing or placement as much as actual airborne concentrations of the pesticide, particularly for pesticides that break down rapidly in the environment (such as organophosphates, whose low environmental persistence belies their high toxicity), or whose use varies widely across space and time (which characterize most pesticides). For example, DPR's ambient monitoring of 1,3-dichloropropene in 2005 as well as methyl bromide in 2005 and 2006 all missed the actual peak use periods, thereby underestimating exposure risks.[53]

Other errors can be made. Samples that are taken for a long period of time do not give an accurate snapshot of short-term spikes in pesticide concentrations. Atmospheric conditions (like wind) could be atypical on the day of sampling, leading to results that underestimate exposure. Various sampling technologies also have their own detection limits; those that are less sensitive than others may not identify pesticide concentrations that are relatively low but still exceed a current health benchmark. Samples often must be thrown away because of equipment malfunctions or human error. Additionally, air-monitoring results for a given pesticide become obsolete when the use of that chemical increases significantly (as has occurred with 1–3,dichloropropene and chloropicrin in recent years) or when application methods change substantially. Hence, at the same time that monitoring studies frequently show that ambient concentrations of pesticides exceed health benchmarks, the various limitations of air-monitoring studies mean that the results may seriously underestimate the degree of pesticide drift.

Health Benchmarks

Thus far, I have presented an array of scientific evidence that pesticide drift often exceeds levels of health concern and have emphasized that the existing data underestimate the scope of the problem. That said, researchers actually disagree considerably about what those levels of health concern—health benchmarks—are and how they should be determined. It is worth understanding the basic dimensions of these disagreements, since the relationship between the health benchmark and the air-monitoring results (i.e., whether the levels in the air exceed a level of public health significance) is what is ultimately of concern to the public and regulatory decision makers regarding pesticide drift. In this section, I describe how health benchmarks are set, what some of the primary points of debate are, and the associated implications for public health.

The health benchmarks I discuss refer to the ambient concentration of a given pesticide that is unlikely to result in adverse health effects. It can therefore be understood as an estimated acceptable concentration of pesticides in the air. Though I use the term health benchmark, researchers and institutions use a variety of analogous terms that will occasionally appear in this book, including "reference exposure level," "reference concentration," "level of concern," and "screening level."

To determine the health effects associated with different levels of exposure to a given pesticide, researchers usually utilize the findings from controlled laboratory experiments on animals. These toxicological experiments are designed to determine the highest dose that results in no observed statistically or biologically significant health effects. Researchers then make a series of mathematical extrapolations to establish a concentration of that pesticide in air (or water) that would result in an acceptably low risk of adverse health effects in human bodies. This extrapolation process involves accounting for numerous points of scientific uncertainty. Notably, risk assessors typically factor in two "uncertainty factors": an *inter*species uncertainty factor to account for the unknown differences between the laboratory animal and humans (such as in absorption and metabolic rates), and an *intra*species uncertainty factor to account for differences among human beings. Each of these is usually factored in as a tenfold margin of safety unless the available evidence suggests that a different factor would be appropriate. Risk assessors occasionally include an additional margin of safety to account for major limitations in the data set; for instance, some risk assessments include an extra tenfold margin of safety for route-to-route extrapolations, such as when dietary study data are used to estimate the consequences of exposure via inhalation. Typically, researchers identify health benchmarks for different possible durations of exposure, such as acute (one to twenty-four hours), short term (one to thirty days), intermediate term (one to six months; also generally called "subchronic"), and chronic exposure (over six months).

After the health benchmarks are determined, they are used to estimate the health risks posed by the concentrations detected in the air-monitoring studies and other exposure studies.[54] Acute or short-term benchmarks generally are used to interpret the results of application site air-monitoring studies. Subchronic benchmarks are used to evaluate the results of ambient air-monitoring studies; researchers look at the average concentration detected per day over the course of one month. This entire process of setting health benchmarks and comparing them with air-monitoring (and other) data is called *risk assessment*, which is a process of first identifying a possible hazard, and then quantifying the probability of exposure (called the "exposure assessment") and the health impacts of various exposure scenarios (called the "dose-response" portion of the analysis).

As numerous scholars have illustrated, the process of risk assessment and the assumptions on which it is based are technically limited and controversial in numerous ways. It is worth taking a moment to consider the various uncertainties, assumptions, and norms that play a role in risk assessments; how risk assessors tend to resolve their conflicts and data gaps; and how the process in turn shapes the regulation of pesticides and use of chemicals in the environment.

Researchers disagree about many key components of the process of setting health benchmarks. For example, there are many debates about how laboratory studies of animals should be conducted. These include questions such as, What durations of exposure best replicate human exposures: how many hours per day, and over how many days? Which "end points" (health effects) to monitor or watch for? Which animals to use? Disagreements among scientists over the design of a study raise questions about whether and how it should be allowed to inform the setting of the health benchmarks and regulations.

Researchers also disagree tremendously about how the toxicology results should be interpreted. Accounting for interspecies variability between animals and humans involves numerous assumptions—for instance, how to account for the fact that animal studies probably cannot detect significant effects that are not readily observable or quantifiable, such as pain, discomfort, and anxiety; and how to account for the fact that short-lived laboratory animals (like mice) simply are not subject to the same range of long-term diseases (like many cancers) that humans are. Accounting for intraspecies variability among humans presents its own set of challenges: How much variation exists among humans in terms of their sensitivity to exposure? How can one determine "average" human breathing rates across a twenty-four-hour period, given that human activity levels vary widely? Do we presume that individuals may be exposed twenty-four hours per day, or that their exposure is zero for part of that time? Should health benchmarks represent acceptably low levels of risk for average adults, or should they include extra protections to account for the special sensitivity that pesticides pose to children, infants, the elderly, people living close to agricultural fields or in poorly insulated homes, people with compromised immune systems, and other special populations?[55]

As a result of all these numerous points of contestation and judgment, there is no singular health benchmark against which any monitoring study results should be compared to determine whether the current pesticide use patterns pose an undue risk to public health. Each institution determines its own health benchmarks for a given pesticide, and the benchmarks can vary widely (for example, see table 2.2).

Numerous scholars have voiced strident concerns about the technical limitations inherent to risk assessment and the extent to which they inhibit the ability of risk assessments to adequately estimate the risks that chemicals pose to human health.[56] First and foremost, the predictive capacity of risk assessment is fundamentally and inherently constrained by the fact that it is used to study singular chemicals. As Steve Wing and others have argued, this reductionist approach—the focus on one chemical in order to isolate its effects, the hallmark of "modern science" that tends to view context as "a nuisance to be avoided by design or controlled by analysis"—ignores the fact that the response of a human (or other animal) to a chemical is contingent on a range of contextual factors.[57] Although health benchmarks represent the health risks posed by exposure to a single chemical, pesticides are typically used in formulations (mixed with other chemicals). Several studies have shown that exposure to pesticide combinations can generate illness effects that are not observed in exposures to the individual, separate pesticides.[58] Little toxicological testing has been conducted on mixtures, so the synergistic and additive effects of such combinations are generally unknown. Similarly, people in agricultural areas are exposed to numerous pesticides over the course of a month, week, or even day. As PAN has often pointed out in its

Table 2.2
Variability in inhalation health benchmarks for adults for acute (24-hour) exposure to chloropicrin

EPA	DPR	EU
491	13	8.56

Note: Numbers represent ambient concentration (measured in terms of $\mu g/m^3$).
Sources: U.S. EPA 2009b; DPR 2010c; European Union 2006.

materials, the air in many agricultural communities is inundated with many pesticides. DPR's PUR data from 2000, for example, show that in the area immediately surrounding and containing the town of Earlimart (the site of a large-scale pesticide drift incident described in chapter 1), agricultural pesticide applications took place nearly every day of the year, averaged 34 per day, and reached as high as 174 in a single day.[59] As a result of these issues noted here, health benchmarks do not account for the additive or combined risks posed by actual exposure scenarios—that is, to many different pesticides in various formulations—and thus routinely underestimate the risk to human health.

Second, risk assessments are widely compromised by serious data gaps. Some major ways in which pesticides likely affect the human body—notably, endocrine system disruption, immune system suppression, impaired neurological development, and for many pesticides, cancer—have been researched very little. Those effects are therefore either not included in the process of setting health benchmarks, or in some cases they are roughly accounted for in the form of a generic uncertainty factor. Most "inert" ingredients (those designed to help distribute or adhere the actual active ingredient) are subject only to minimal testing, and manufacturers are not required to disclose the inert ingredients contained in pesticide formulations because they are considered confidential, proprietary business information.[60] Moreover, many risk assessments for fumigants and other drift-prone pesticides do not include sufficient direct data from inhalation laboratory studies, since toxicological research of that type is rarely conducted. In such cases, researchers must rely on dietary or dermal exposure studies, and estimate the relationship between those other exposure route data and the risks posed by inhalation. This dearth of inhalation data is no small matter. Many health benchmarks determined directly from inhalation studies are much lower than those derived from studies of other exposure routes.[61]

Additionally, most risk assessments do not include data on the health effects of chronic exposure to low levels of pesticides—which is precisely how most agricultural community residents experience pesticide exposure.[62] The health effects of such exposures may not even be accurately determined from standard toxicological methods that assume a positive relationship between dose and response ("the dose makes the poison"). Scientists who have recently begun to study the consequences of exposure to low doses of certain chemicals have found that some dose-response curves defy this assumed relationship and cause illness at low levels— that is, below the levels previously thought to be benign. Pesticides that disrupt the human endocrine system exhibit this characteristic; for such compounds, the response is not determined so much by the dose as by the timing of exposure and the age of the individual.[63] This indicates that many previously determined health benchmarks do not adequately protect human health.

Third, risk assessments are also limited by standard scientific norms. Widely accepted scientific norms, for instance, dictate that when some uncertainty exists in terms of a relationship between two variables (say, a pesticide and a disease), the researcher should err on the side of assuming that the relationship does not exist even though in fact it might, rather than assuming that the relationship *does* exist when in fact it might not. In other words, because of such norms, scientists will generally assert that a chemical causes a certain health outcome *only* if they are absolutely certain.[64] While this sounds reasonable in theory, it works against health-protective measures in practice. Where a study contains a small number of data points and a high degree of variation—as characterizes much toxicological research—it may be impossible to draw a definitive conclusion about the relationship between a pesticide and illness. This is particularly the case at low levels of exposure, where health effects are less obvious and hence more difficult to detect.[65]

Simply put, there is a lot of information that scientists do not know, and the predominant scientific norms described above compel scientists to assume a risk is zero until sufficient evidence suggests otherwise. The assumptions on which risk assessment are based—namely, that we can anticipate the planet's capacity to absorb toxins, can accurately set regulations so that toxic loads will not exceed that capacity, and already know (or can easily determine) the harms posed by any given chemical—therefore are simply untenable for situations in which key pieces of scientific data are missing. This unfortunately applies to nearly every agricultural pesticide in use today.[66]

In sum, the available illness data and air-monitoring studies document pesticide drift of highly toxic pesticides at levels known to cause ill effects in animals. More troubling is the fact that these data *under*estimate the incidence of pesticide drift and drift-related illness, on account of the rarity of air-monitoring studies, the innumerable limitations of both illness and monitoring data, and the problems with the way many health benchmarks are typically set. As a result of the limitations of air-monitoring and verified pesticide illness data, understanding the *likely* consequences of pesticide drift will require that we consider a wider array of information, as I discuss in the next section.

Human Health Consequences of Pesticide Drift

Illness and air-monitoring data show that pesticide drift occurs frequently at levels that exceed health benchmarks. Moreover, those data sets and the ways in which health benchmarks are set are limited to such an extent that they underrepresent the actual incidence of pesticide drift, exposure, and illness. Understanding the human health consequences of pesticide drift, then, will require that we examine indirect and informal data sets about pesticide illness.

In addition to the PISP data on acute illness discussed above, a wide array of evidence suggests that the consequences of pesticide drift are profound and pervasive. For example, researchers from California's Department of Public Health estimated the risks of harmful exposure to agricultural pesticide drift facing communities throughout the state by combining ambient air-monitoring, pesticide use, and census population data. The authors concluded that hundreds of thousands of Californians every year are exposed to levels of metam sodium/MITC, methyl bromide, 1,3-dichloropropene, and chlorpyrifos that exceed noncancer health benchmarks, and levels of 1,3-dichloropropene, methidathion, and molinate that exceed cancer health benchmarks.[67] The authors note that actual exposures are likely much higher, given that their study only addressed inhalation exposure (whereas people, especially children, are commonly exposed via dermal and dietary exposure as well) and only used ambient air-monitoring studies (instead of data from application sites, where concentrations are higher). The authors noted also that exposures may be higher in other states whose pesticide use regulations are less stringent than California's.

Additional evidence suggests that exposures to pesticide drift cause serious chronic illness. For example, residents involved in documented cases of pesticide exposure report a litany of ongoing chronic health problems that they attribute to their exposure to pesticide drift: asthma, migraines, rashes, nosebleeds, chronic bronchitis, damaged eyesight, miscarriages, heart attacks, learning disabilities, and behavioral changes. Many pesticide drift exposure victims also emphasize the lasting emotional effects of the physical trauma as well as the neglect and mistreatment by emergency crews, noting depression, poor self-esteem, and fear.[68] In his research on community exposure to toxic waste, Phil Brown has illustrated the lasting and profound emotional trauma experienced by toxic exposure victims and their family members.[69] Despite the serious nature of these lasting health impacts, regulators rarely investigate, document, or follow up on reports of the long-term physical and emotional consequences of pesticide drift exposure.[70]

Epidemiological research takes this discussion of individual experiences one step further by correlating the use patterns of particular pesticides with the prevalence of disease and disability likely to result from those pesticides. Epidemiologists and other researchers have found that people living in areas of high use rates of certain pesticides are subject to a higher risk of numerous forms of cancer, birth defects, asthma and other respiratory disorders, miscarriage, low birth weight, stillbirth, infant mortality, infertility or reduced fertility, learning disorders, autism spectrum disorders, and Parkinson's disease.[71] Wildlife epidemiologists provide ample evidence that chronic exposure to low levels of pesticides (i.e., even below regulatory agencies' health benchmarks) suppresses immune system function, rendering people and other animals less able to combat disease. In recent years, wildlife researchers have correlated agricultural pesticide use with novel and devastating recent disease outbreaks among honeybees, bats, and amphibians, and there is ample evidence to suggest that these chemicals have similar effects on the human body.[72] This staggering array of evidence notwithstanding, epidemiological research is best understood as illuminating the tip of the iceberg in terms of pesticides' contribution to disease and suffering. Epidemiology alone simply cannot provide conclusive evidence of cause and effect between individual chemicals and health outcomes, given the wide array of confounding variables (i.e., other factors that contribute to disease).[73]

Conclusion

When examined in its broader social context, scientific research provides an ominous view into the physical, material world of pesticide drift. Illness data and air-monitoring studies demonstrate that pesticide drift is pervasive, often exceeding health benchmarks. Moreover, these technical data sets and health benchmarks underrepresent risk and exposure because of various reasons: the technical limitations of scientific research, the constraints of scientific norms regarding uncertainty, the profoundly complex and dynamic nature of agricultural pesticides and the physical environment through which they move, and multiple deeply rooted social inequalities. Consequently, pesticide drift deeply shapes the lives of millions of people in California's agricultural communities—contributing to long-term illness, psychological trauma, and reduced quality of life—yet it remains largely unrecognized. Viewed in light of the available scientific evidence as well as a nuanced understanding of the limits of that science, pesticide drift emerges as a systematic though invisible example of widespread chemical contamination.

This is not to say that institutions and individuals do not struggle to control the movement of agricultural pesticides into air, water, and bodies. Indeed, California's progressive reputation accurately applies in many ways to pesticide drift as much as it does to other environmental issues. The question is not whether actions have been taken to address the environmental and public health impacts of California's agricultural pesticide use—since, as I will show, many have. Rather, our task is to examine what those actions have been and how effectively they control pesticide drift. In the next three chapters, I discuss the measures taken by the agricultural industry, environmental regulatory agencies, and alternative agrifood activists to tackle the extraordinary environmental and public health impacts of agricultural pesticide use in California. I will show that their efforts and accomplishments, while numerous and laudatory, are fundamentally disabled not simply by the limits of science and inadequate technologies but also by profoundly unequal social relations and a problematic conception of social justice.

3

The Crop Protection Industry

The social responsibility of business is to increase its profits.
—Milton Friedman, *New York Times Magazine*

The accumulated hazard of chemical production and consumption is not a conspiracy dreamed up in a cigar smoke-filled back room. Rather, it is one of countless unintended but logical outcomes of larger forces, consolidations, and negotiations of global industries in early twenty-first century capitalism.
—Paul Robbins, *Lawn People*

Introduction to the Industry of California Agriculture

Countless writers have described California's pivotal role in the emergence of the highly specialized form of growing food and fiber that we refer to as conventional, modern, or industrial agriculture.[1] As historian Steven Stoll points out, the exceptionally productive nature of California agriculture does not stem as much from any sort of "natural advantage" as it does from the deliberate work of many investors, researchers, and political leaders.[2] California agriculture underwent particularly significant changes in the 1880s, as capitalist investors transformed the state's agricultural industry from one dominated by wheat to a powerhouse of high-value, resource-intensive, specialty fruit production. These investors and growers drew on many technological and institutional innovations to make export-oriented specialty fruit production possible as well as exceptionally profitable. The transcontinental railroad and refrigeration technologies enabled the shipment of otherwise quite perishable fruits to distant markets, and the development of the state's irrigation systems enabled farms to prosper in what is generally a rather arid region. Agricultural investors drew on the horticultural knowledge and resources of the many immigrant groups that had settled in the state, including the different fruit varieties they introduced and the expertise needed to grow them well. Growers formed cooperative marketing associations, designed and implemented commodity quality grades and standards, and constructed packinghouses to meet consumer expectations and get their products to market quickly. Because specialized, large-scale agriculture often requires large numbers of workers for short periods of time during the year (e.g., at harvest time), California growers enlisted the role of the state in recruiting immigrant workers, suppressing their rights and wages, and compelling them to leave when the work was done.[3] Additionally, as will become clear in chapter 4, the state's particular approaches to designing and enforcing environmental policies and regulations further nurtured the development of industrial agriculture.

None of these innovations was more important than the emergence and development of the chemical pest control (or in industry parlance, "crop protection") industry. Throughout the nineteenth century, California growers struggled with a wide variety of insect pests and plant diseases, many of which had entered the state by hiding on imported, exotic varieties of fruits. Pests' subsequent proliferation in California's nurturing climate and wide expanses of monocropped fields destroyed many capitalist farmers' investments. Because each grower had already invested tremendous resources into expensive systems focused on the production of a singular crop, they generally rejected the approach of combating pest populations through diversifying their agroecosystems (which can encourage natural predator insects and otherwise help to disrupt pests). Instead, growers embraced chemical technologies as the primary approach to controlling pests. As fields fell under attack by insects and other pests, an entire research apparatus and industry emerged to solve these problems. This specific chemical control industry would come to have grave consequences for life in and near agricultural fields.

This chapter examines the agricultural pest control industry to help illustrate how toxic chemicals have come to be used so widely that they seem to be a natural part of the agricultural landscape, and how in turn, pesticide drift came to be a pervasive problem. As this chapter will show, the "crop protection industry" is now an elaborate, global, and complex network of institutions and individuals. I focus my discussion here on those actors who are the most invested in agricultural pesticide use and who most directly shape pesticide use, drift, and illness: pesticide manufacturers, distributors, and end users as well as the organizations that represent their collective interests. This diverse collection of institutions and individuals can be understood as an interactive and mutually supportive network of relationships governed by a shared commitment to the pesticide paradigm.

As will become clear, industry actors have made some notable changes in the name of environmental sustainability that could arguably help to reduce the problem of pesticide drift. Yet those efforts are ultimately fundamentally constrained by industry's imperative to maximize profit. The most economically successful pesticide industry actors achieve this maxim through promoting existing pesticides, distorting scientific debates, and fighting environmental regulations and legislation. Of course, such practices pose extraordinary environmental and public health costs, but industry consistently shirks its responsibility for such costs by shifting them elsewhere: to the bodies of people and other animals living downstream and downwind as well as to those who have not yet been born.

Chemical Manufacturers: Developing and Creating Pesticides

Growers in California and elsewhere had long experimented with various toxic compounds and mixtures to kill agricultural pests, often with little effectiveness and generally with a poor understanding of insect physiology. In the late nineteenth century, researchers from the University of California helped to develop effective chemical pesticides and application practices.[4] The model of pest management that came to dominate research and practice was one that accommodated, rather than challenged, the same industrialized, ecologically simplified model of agricultural production that had generated massive pest problems in the first place. Though a contingent of University of California entomologists during the same time ardently promoted biological forms of pest control, expressed concern about the temporary nature of chemical pesticides' effectiveness, and warned of their toxic effects on beneficial insects, university involvement in agricultural pest control quickly coalesced around chemical pesticides. University researchers helped to develop powerful chemical pest controls, and their credibility as public servants was a welcome relief to farmers, given the overabundance of fraudulent pesticides on the market. Two University of California researchers started their own pesticide manufacturing company in 1907: California Spray-Chemical, the first such enterprise springing from university-based scientific research. Stoll emphasizes that California Spray-Chemical was innovative for making "scientific research the engine of market innovation"; the company advertised itself as "scientific pest control" and extolled its products as "Correct, Proper, Right, Straight and Pure."[5] University leaders actively promoted the commercialization of pest control technologies, welcoming industry's ability to disseminate innovations and assuming that it would do so responsibly.[6]

The U.S. chemical industry flourished rapidly during World War I and World War II, largely through the development of chemical technologies for warfare that were later adapted to other purposes. Historian Edmund Russell shows how the growth of the chemical industry cannot be understood apart from its role for wartime; chemical warfare and agricultural pesticides coevolved ideologically, technologically, and organizationally, and in the process helped to reinforce and legitimize the industrialized model of agriculture.[7] During the world wars, the U.S. chemical industry produced explosives, poison gases, dyes, and pesticides on government military contracts. Through these wartime accelerations in research, chemical companies discovered that some by-products of dye and explosives manufacturing had powerful pesticidal properties, and also created entirely new classes of synthetic chemical pesticides expressly for use on humans and insect pests (including organochlorines and organophosphates). After the war effort, chemical companies successfully adapted many pesticides to agricultural uses. The U.S. government directly encouraged this process through various means, including directing and funding the development of aerial pesticide application technologies along with conducting large-scale campaigns to eradicate public health pests with chemical controls. Government contracts for chemical industry products during wartime, in short, facilitated the development of the U.S. chemical industry, which then used its accumulated capital, skills, and continued government contracts to expand into new markets—notably, home and agricultural pest control.

The chemical manufacturing industry has continued to grow in size over time. The global agricultural chemicals market was valued at $44 billion in 2007 and is expected to reach a value of $80 billion by 2012.[8] Additionally, pesticide manufacturing has become an extremely concentrated sector, in which only the most highly capitalized firms can compete.[9] Nearly three-quarters of all pesticides sold globally in 2004 came from just six multinational corporations: Bayer, Syngenta, BASF, Dow, Monsanto, and DuPont.[10]

Going "Green"
Though these corporations have made pesticide development and manufacturing a lucrative endeavor, many conditions threaten manufacturers' investments. Insects continually develop resistance to specific pesticides—a process of biological adaptation that has rendered many pesticides ineffective over time. Also, when the patent period for a pesticide closes, as

with any investment, the original manufacturer must compete with less expensive, generic pesticides. Public concern about the environmental and public health impacts of pesticides has increased steadily in the past forty years, too, thus prompting a demand for less toxic pesticides as well as new environmental regulatory controls on many existing pesticides. The chemical industry's legitimacy had declined in the 1980s in response not just to the rise of environmentalism but also to several major industrial catastrophes—notably, a major chemical leak from a Union Carbide factory in Bhopal, India, in 1984 that killed tens of thousands of residents and caused thousands to continue to suffer from lasting chronic illnesses and injuries.[11]

Facing these challenges to their existing investments and reputation, pesticide manufacturers continually develop new products and nurture new markets. Many of these changes in recent years have been done in the name of environmental sustainability. For example, pesticide manufacturers have developed pesticides with lower environmental toxicity or have partnered with smaller companies that have developed such products. A market that is currently valued at a billion dollars, industry leaders are investing heavily in less toxic pesticides and expect the market to grow to $5 to $10 billion worldwide by 2017.[12] Manufacturers also produce and market compounds that can be added to pesticide applications to reportedly reduce the drift potential of pesticide sprays. Genetically modified (GM) seeds constitute another major technological innovation that agricultural chemical manufacturers have pursued in the name of environmental sustainability. In fact, the top pesticide companies control the majority of the GM seed market. Manufacturers promote many agricultural biotechnologies as environmentally friendly ways to reduce pesticide use (such as crops that have been engineered to exhibit insecticidal properties).[13]

In addition to these technological innovations, all of the major pesticide manufacturers participate in a variety of private, voluntary initiatives designed to promote environmental sustainability through pesticide applicator training, sustainability reporting, and industry self-regulation agreements. Industry representatives encourage education efforts to train pesticide applicators in proper pesticide application practices. For example, the National Coalition on Drift Minimization in the United States brings together industry, regulatory, and university representatives working to minimize drift through developing training materials for pesticide applicators and advising pesticide regulators.[14]

Chemical manufacturers promote environmental sustainability through participating in corporate sustainability reporting. All major pesticide manufacturers, for instance, prepare their own self-assessed sustainability reports in accordance with the UN-sponsored Global Reporting Initiative guidelines. These guidelines constitute a framework that companies use to evaluate their own performance in terms of a range of economic, social, and environmental sustainability criteria. Moreover, as they are eager to point out, most of the major pesticide manufacturers have been named to the Dow Jones Sustainability Index (DJSI), which tracks the financial performance of what it has determined to be "the leading sustainability-driven companies worldwide."[15] In fact, Syngenta was recently awarded "gold status" in the DJSI, placing it in the "top tier" of all DJSI-ranked companies.

Chemical manufacturers also participate in industry self-regulation schemes. The most notable example is that most major pesticide manu-facturers helped to develop and continue to endorse Responsible Care, the chemical industry's own voluntary sustainability self-regulation program. In Responsible Care, participating chemical companies commit to a set of key environmental, health, and safety principles, to which they can evalu-ate their own adherence via a set of measurable indicators.[16]

There are good reasons to be wary of sustainability reporting and self-regulation as a means to reward and encourage sustainability, however. In sustainability reporting and ranking schemes, environmental sustainability indicators comprise only a small portion of all the various sustainability criteria. The DJSI, for example, places considerably greater emphasis on economic criteria (including "corporate governance" and "risk manage-ment") and social criteria (including "talent attraction and retention") than it does on environmental criteria.[17] Sustainability assessments also rely on a limited range of ecological indicators. For example, the Keystone Alliance's sustainability assessments explicitly ignore the harmful effects of agricultural chemicals except for their implications for "energy use" and "climate impact."[18] Responsible Care, although touted by U.S. EPA officials as the most advanced industry self-regulatory program, does not specify the extent to which participants must adhere to its guidelines, provides no third party to verify adherence to those guidelines, and con-tains no sanctions for participants who inadequately implement those guidelines.[19]

Many observers regard industry self-regulation as a fundamentally com-promised approach to addressing many environmental problems. Their voluntary basis means that participation and the associated environmen-tal benefits are uneven. They also are typically characterized by limited accountability and reliability, since few contain mechanisms to verify participants' biased self-assessments or sanctions to prevent opportunistic firms from free riding on other participants' efforts.[20] Such problems are an inevitable weakness of industry self-regulation programs, since incorporating sanctions would undermine the "voluntary" and industry-friendly quality that compels firms to participate in those programs in the first place.[21] Indeed, industry participants have vociferously defended the voluntary and unenforceable nature of such programs.[22] In a world characterized by major imbalances of power, environmental advocates are relatively unable to compel industry to set strong and enforceable sustainability standards.[23] Finally, it should be recognized that some limitations of chemical manufacturers' voluntary sustainability efforts are due to the nature of the industry itself. Chemical manufacturers share an incentive to protect the perception/reputation of the industry—not to reduce pollution per se. In contrast to other industries whose participants have a strong economic incentive to manage their shared resources at ecologically sustainable levels (e.g., international fisheries), agricultural pollution does not compromise chemical manufacturers' profits.[24] It is for these reasons that many analysts have characterized such industry efforts as elaborate public relations campaigns that have little environmental consequence.

Some of the agrichemical industry's so-called green efforts have decidedly negative environmental impacts as well. Herbicide-tolerant GM crops have enabled farmers to till less frequently but also have created herbicide-resistant "super weeds," to which farmers respond by applying additional types of herbicides, applying them more frequently, and increasing the rates at which they apply them. Also, insecticide-resistant GM crops have led to increased insecticide use, since these crops kill beneficial insects and increase pest insects' resistance to the toxins engineered into the plant. Such problems are not inconsequential: the majority of all corn, soy, and cotton acreage in the United States is now planted with GM seeds (99 percent of which contain properties of herbicide tolerance or insect resistance).[25]

Protecting Existing Investments

At the same time that chemical manufacturers' green efforts are proliferating, manufacturers have both the motivation and resources needed to defend their existing investments and markets—that is, to promote the sales and use of all their products, including the most toxic ones. Manufacturers are compelled to aggressively market their existing products on account of numerous factors that make the agricultural chemical market a risky one: the costly and time-consuming nature of research and development, the costs of regulations, patent rotations that allow the introduction of cheaper, generic pesticides, pest resistance to particular chemicals, and the variability of demand between seasons and from one year to the next. In other words, the same factors that prompt manufacturers to invest in less toxic technologies to fill new market niches also compel them to stridently defend as well as expand the markets for their old products, shift risk, and actively depict pesticides and their users in a positive light. Moreover, they have the economic resources needed to do so.

This is no more evident than in current debates over agricultural soil fumigants, widely regarded as the most drift-prone and toxic group of agricultural pesticides in use today. At the same time that chemical manufacturers promote their sustainability claims, they stridently lobby against new health-protective regulations for the existing soil fumigants (methyl bromide, metam sodium, metam potassium, chloropicrin, and 1,3-dichloropropene) and for the registration of two new fumigants for use in agricultural settings (sulfuryl fluoride and methyl iodide)—in spite of the fact that the latter two fumigants are widely regarded in the scientific community as exceptionally problematic for public health and contribute to global warming.[26]

Industry protects its existing investments through a variety of mechanisms, including expanding into new markets, shaping scientific and public debate, influencing regulators and policymakers, violating environmental and labor protections, and carefully crafting a green image in advertising campaigns. Chemical manufacturers are always in search of new markets, and in recent years have expanded into new regions such as Latin America, especially to market the oldest, most hazardous pesticides that are being increasingly regulated in the United States and European Union.[27] Also, Paul Robbins has shown how chemical manufacturers, seeing their agricultural chemical profits slowing, have aggressively expanded the lawn chemical market in the United States.[28] Many of the chemical manufacturers' green technologies and practices are also specifically designed to permit a greater use of pesticides (as is the case with the herbicide-tolerant crops that are explicitly designed to withstand high rates of herbicide applications). In addition, chemical manufacturers have widely promoted herbicide-intensive "no-till" (or conservation tillage) agriculture, where herbicides are used to replace tillage. Promoted as a green practice because of its ability to help farmers reduce soil erosion, those environmental benefits are complicated by the fact that its rapid growth in the United States and Latin America in recent years has driven up the sales and use of herbicides.[29]

Although the agrichemical industry argues that its economic growth is a response to increasing pest pressures, the companies spend millions of dollars per year promoting routine ("preemptive") pesticide applications and otherwise actively constructing that demand in order to sell more product.[30] As in any industry, pesticide advertisements deliberately draw on current social norms to make their products appealing to consumers. Sociologists, for instance, have found that pesticide advertising drew largely on "better living through science" narratives in the chemical revolution of the 1950s and 1960s, shifted toward narratives of "control" during the cold war years of the late 1960s and 1970s, and increasingly pulled from environmental narratives since the rise of public concerns about the environment in the 1980s.[31] Despite the rise of environmental framings, current pesticide advertisements continue to actively promote an ecologically simplified model of agriculture in which advanced scientific technologies (e.g., the latest pesticides) dominate and control unruly natural elements in orderly fields. Advertisements are replete with images of hunting, other combat-based sports, and warfare, which are superimposed over "clean" fields free from plant and insect pests. As Robbins's work on lawn chemical advertising suggests, pesticide advertisements construct not just good crops but also the image of a "good farmer"—one who is concerned about the environment, yet principally in charge of his field and in control over the nature within it.[32] My choice of pronouns here is no mistake; as illustrated in figure 3.1, below, pesticide advertisements directly appeal to as well as reinforce dominant heterosexual masculine norms (in this case, directly linking pesticide use with fatherly responsibility, sports, and uniformly green outdoor spaces). A 2007 Bayer pesticide ad highlighted related norms, boasting "Serious Vegetable Seed Protection Up and Down the Field" over an image of a sweaty, muscular, growling, larger-than-life football player wearing a red, white, and blue uniform and crouched over a field of immaculate rows of corn. Similarly, a 2008 Syngenta pesticide ad featured a photograph of a pair of shapely (and pale-skinned) woman's legs in a short skirt and high heels, foregrounded by a plate a colorful vegetables, announcing that "Beautiful vegetables and potatoes change the way you look at things."

The agrichemical industry also protects its existing investments by actively shaping scientific and political debate in its favor. Chemical manufacturers and other major industry groups regularly fund trade organizations, think tanks, and independent experts to conduct research, and they then disseminate the particular results that represent pesticides as benign or beneficial. Critical observers have shown that such studies often contain

Figure 3.1
Pesticide advertisement from DuPont. Copyright © 2010 E. I. du Pont de Nemours and Company. All rights reserved.

egregious and deliberate design errors so that the results will show that industry products are safe despite evidence to the contrary.[33] In other cases, such research is designed to deliberately generate scientific ambiguity and doubt about the risks posed by a particular pesticide; the creation of scientific disagreement suggests "the need for further scientific research" and effectively stalls regulatory action. Industry-funded experts actively attack and discredit scientists whose research identifies the dangers of chemical technologies, as evidenced by the industry's efforts to discredit Rachel Carson and her book *Silent Spring* throughout the 1960s.[34] Industry-funded experts defend company interests in chemical liability lawsuits and write letters to the editor of newspapers in defense of the industry. Such work contains the illusion of impartiality in spite of being paid directly to serve chemical industry interests.[35]

Agrichemical firms also directly intervene in research conducted at public universities, selectively disseminating those results that legitimize industry investments and expand market opportunities while suppressing damning results.[36] Public-private research relationships are increasingly common in an era of declining funds for public education, and private companies now fund upward of 25 percent of all research conducted in some public university departments.[37] Although the actual terms of any given contract determine to a large extent how much control the donor firm has over the research process and results, researchers have shown that scientists receiving industry funding tend to be biased toward industry interests even in cases where the industry sponsor does not actively pressure the researcher.[38]

Another primary mechanism through which chemical manufacturers protect their investments in existing pesticides is by shaping important decisions in the regulatory and policy arenas in their own favor—particularly by dedicating tremendous resources to fighting proposed regulatory restrictions on the use of specific pesticides and otherwise keeping those markets as open as possible. Chemical manufacturers and their trade associations influence regulatory and legislative decisions about pesticides through direct lobbying, making campaign contributions to candidates for federal offices and to political parties, funding ballot initiatives that support their interests, and funding campaigns to defeat ballot measures that threaten their interests. Environmental organizations also engage in these tactics, but their meager budgets are greatly overwhelmed by those of the major agricultural chemical manufacturers.[39]

Pesticide manufacturers hire professional lobbying firms to do much of this work, as such firms provide special access to policymakers and regulatory officials because of their employees' past work experiences within the policy and regulatory arenas. As of late 2009, for example, 32 currently registered lobbyists working at the federal level had been on or worked for the Senate Agriculture, Nutrition, and Forestry Committee; 50 had been on or worked for the House Agriculture Committee; 89 had past positions in the U.S. Department of Agriculture (USDA); and 114 had worked at the EPA.[40] Observers often critically characterize these types of relationships between industry and government as a revolving door through which individuals in leadership positions rotate between industry and the public agencies that regulate them, and provide "insider know-how and friendly connections through which rules can be bent and loopholes exploited."[41]

Industry also influences policymakers by creating and funding "front groups" that are designed to look like grassroots collections of citizens or associations of professionals, but are actually funded and controlled entirely by industry.[42] One of the most recent such group to emerge is the Coalition for Chemical Safety, comprised of major chemical corporations and trade associations, yet self-described as "people like you working to ensure a balanced approach to our nation's chemical safety laws . . . a non-profit social welfare organization committed to protecting families and communities today while encouraging the development of safe green chemicals for tomorrow."[43]

Like all other corporations, chemical manufacturers also protect and bolster their profits by extracting as much value as possible from labor and the environment. Environmental damage and social exploitation are thus a hallmark of capitalist business practice, even in an era of increased environmental concern and knowledge. This is particularly the case when the benefits of violating labor and environmental regulations are greater than the associated fines or other sanctions, as is true throughout much of the world. It is for these reasons that chemical manufacturers have been found to have committed innumerable deplorable acts, including knowingly dumping toxic wastes and obsolete chemicals, violating human subjects testing protocol, violating worker health and safety protections, busting labor unions, misrepresenting their products through false advertising, producing and distributing globally products that have been banned in the United States, and failing to take responsibility for chemical accidents.[44]

In sum, pesticide manufacturing is a highly consolidated industry dominated by six major multinational corporations. In response to the various threats to their investments, manufacturers have expanded their investments in green pest management technologies at the same time that they aggressively protect and expand the markets for their older, more dangerous pesticides. It is for these reasons that pesticide manufacturers' self-congratulatory forays into environmental sustainability are best understood as much for what they conceal as for the environmental gains they accomplish.

Pesticide Distributors: Advice, Sales, Delivery, and Application

Pesticide distributors serve as one-stop agrichemical shopping for farmers, providing advice, selling pesticides and fertilizers, delivering those products to the users, and in some cases, applying the products. Although physically dispersed widely across the agricultural landscape in the form of thousands of relatively small retail centers, pesticide distribution is a highly concentrated industry in the United States. The top five agrichemical distributors include Agrium (a recent merger of Western Farm Service, Crop Protection Services, and United Agri Products), Helena, Wilbur Ellis, Simplot, and Growmark. Sales from these top five multinational corporations amounted to $11.8 billion in 2008, accounting for over 60 percent of the total sales from the top one hundred distributors.[45]

In addition to delivering and selling pesticides, distributors are also the most important source of pest management advice for California growers. Distributors' salespeople are certified pest control advisers (PCAs; they are known as certified crop consultants throughout the rest of the United States). PCAs are the only individuals authorized to make specific pesticide recommendations in the state of California, and nearly 90 percent of all PCAs work for major pesticide distributors and earn a substantial commission on all pesticides sold. These individuals are growers' primary source of advice about pesticides. Within the range of a handful of distributors in any particular agricultural area, a grower will select a distributor and PCA who they trust, feel comfortable with, and believe has the experience and services they need.

Pesticide distributors emphasize their environmental sustainability efforts. For example, Agrium's latest sustainability report boasts that the company helps to develop precision farming techniques, sells genetically modified seeds that may reduce pesticide use, promotes partnerships with various stakeholders "to encourage the responsible management and use of our products and services," and adheres to existing laws and regulations.[46] The California Association of PCAs (CAPCA) works to address problems like pesticide drift largely through overseeing continuing education programs for certified PCAs. Additionally, CAPCA asserts that the highly competitive nature of the market within which PCAs vie for clients itself functions to protect the environment, because overprescribing chemicals could cause one PCA to lose his or her account to another PCA who could serve the grower's needs more economically.[47]

Despite these claims, many material factors hobble pesticide distributors' contributions to environmental sustainability and leave them in a poor position to reduce problems like pesticide drift. Numerous critics have argued that the fact that commercial PCAs' salaries are largely comprised of commissions from pesticide sales constitutes a significant conflict of interest and institutionalizes unsustainable pesticide use practices. Indeed, the occupation was originally more explicit: PCAs used to be simply called pesticide salespeople, but the profession shifted slightly (and largely rhetorically) when political action in the early 1970s required them to take an exam and maintain a PCA license.[48] PCAs have no economic incentive to recommend a form of pest control that does not include selling any product, as is the case with many alternative approaches to pest management (which, instead of applying a commercial product, require increased monitoring of pest populations, rotating crops from one year to the next, planting a wider variety of crops, and providing habitat for pests' natural predators). PCAs' extraordinary conflict of interest has been criticized over the years, yet chemical companies have successfully fought all legislative efforts to separate the institutions of pesticide advice and pesticide sales.[49] CAPCA recognizes that the fact that PCAs earn commissions on pesticide sales poses some conflict of interest, but the organization insists that environmental regulations and the considerable competition among PCAs prevent any significant problem.[50]

This is not to say that all PCAs are handmaidens to chemical manufacturers. Some PCAs work independently of agrichemical companies, charging more for agroecological services (such as monitoring pest populations) than for sales of specific products. Also, some individuals with advanced training in crop science and applied entomology work out of public universities as "farm advisers."[51] Yet like growers, all PCAs err on the side of overprescribing pesticides in order to ensure the crop's success in the immediate term. Keith Warner explains that PCAs perceive ecological, knowledge-intensive alternative pest management practices as too risky:

PCAs know that a wrong decision could push a marginal grower to the brink of economic disaster. . . . A PCA's continued occupation is dependent on his good reputation in the agricultural community, and if more than one bad decision is made, he or she may have to seek work in another field. . . . They know that a bad reputation is much more difficult to repair than a questionable pesticide recommendation.[52]

In this context, there is no incentive to adequately conduct the labor-intensive work essential for reducing pesticide use (such as regular, careful monitoring of pest populations), and PCAs have been widely critiqued for overprescribing pesticides and doing too little monitoring. Even those PCAs who participate in innovative institutional partnerships to reduce pesticide use complain that such projects ask "them to do more and be compensated less," since they still operate in the same context as all PCAs do: needing to help ensure growers' yields.[53]

Many of the specific material structures that constrain the work of PCAs similarly limit the ability of pesticide distribution companies to pursue environmental sustainability. The primary constraint is a simple one: pesticide distributors' profits are contingent on selling products. Ecological approaches to agricultural pest management rarely involve merely substituting one profitable, less toxic pesticide for an older, more toxic one. Rather, they include replacing many chemical inputs with labor-intensive practices. Reforming pest management to reduce problems like pesticide drift, in other words, threatens the market that distributors have crafted, or at the least, requires changes that pesticide distributors cannot easily and profitably provide. As a result, pesticide distributors protect their current investments and market share by working to keep as many products on the market as possible. Distributors therefore participate in, fund, and are represented by many of the same trade associations that represent the interests of pesticide manufacturers: CropLife America, Western Plant Health Association, and Agricultural Retailers Association—those same associations that lobby hard against proposed regulatory restrictions on agricultural pesticides. Additionally, Bayer and DuPont are the two major funders for CAPCA.[54] Selling highly toxic pesticides is an important part of distributors' profits. Thus, it is no surprise that at the same time that pesticide distributor corporations report their sustainability practices, they also promote the unequivocally unsustainable practice of "preventive" pesticide use. As Agrium noted in its 2008 annual report, "We also anticipate strong demand for crop protection products due to higher than average crop prices supporting preventive chemical applications."[55]

Agricultural Pesticide Users: Growers, Professional Applicators, and Hired Workers

The preceding discussion of pesticide manufacturers and distributors illustrates how off-farm industry actors play a role in shaping pesticide use practices, patterns, and impacts. Of course, the individuals most immediately responsible for agricultural pesticide drift are the pesticide users: farmers and the people they employ to apply pesticides on to their crops and fields. Although farmers' demand for agricultural chemicals fluctuates to some extent with commodity prices, weather, pest pressure, and the strength of the agricultural economy, agricultural pesticide use in the United States remains steady over time.[56]

As described earlier in this chapter, California's fruit and vegetable producers have long relied on chemical pesticides as the core (and in many cases, the *only*) set of technologies for combating agricultural pests. Chemical pesticides thus served a crucial role in the early development of industrial California agriculture, as they maintained the productivity of an oversimplified type of agroecosystem vulnerable to pest infestations. Farmers' use of chemical pesticides skyrocketed after World War II because the chemicals that emerged from that time were easy, effective, relatively cheap, and widely encouraged by university extension agents and academic researchers. Although these were new technologies, chemical pesticides soon became popular not just because they appeared to quickly solve pest problems but also because they aligned well with the predominant ideologies of the day. Compared to other pest management systems that require careful monitoring and work best when neighboring landowners cooperate, chemical control methods are quick and do not require collective action. As Christopher Bosso notes, chemical pesticides fit into the cultural structures of farming communities, as they "suited farmers' individualist ethos—each in control of his own destiny." The chemical-intensive model of pest control generally fit well with the predominant social values of the mid-twentieth century: convenience, simplicity, and "better living through science" as well as the predominant understanding of progress as meaning expanding production through controlling nature.[57]

Today, farmers acquire most of their pest management information from PCAs. Each grower generally works with one PCA (typically one who works for a commercial pesticide distributorship and hence earns a commission on all pesticides they sell), who plays a crucial part in identifying pest problems and recommending solutions. Some farmers apply their own pesticides and fertilizers, particularly perennial crop growers, since they will be using the appropriate crop-specific application equipment for many years and thus own that machinery. When growers need to hire someone to apply pesticides, they contract either the distributor or a local professional pesticide application company. Even when farmers do not conduct the actual pesticide applications, they make the final decisions about the pesticide application schedule (the frequency, intensity, and method of application).

Grower organizations (including commodity organizations, grower cooperatives, and trade organizations) play a key role in upholding the centrality of chemical pesticides in California agriculture. Established originally in the early twentieth century with the state's help to facilitate cross-country transportation, marketing, the development of quality standards, and the conduct of research, grower organizations are a primary source of information for growers about pest management practices and regulatory changes. These organizations currently represent farmers' interests in the public and political arenas through various means, including enhancing demand for farm commodities in the marketplace, shaping university scientists' research priorities through selective financial and political support, lobbying on proposed pesticide regulations, and disseminating scientific research findings that support their constituents' interests.

Farming More Sustainably

In recent years, many growers have proactively engaged in a variety of voluntary efforts to reduce the environmental impacts of their farming operations. For example, some participate in "grower partnerships" that bring farmers together to help each other figure out how to reduce their pesticide use and other negative environmental impacts. Researchers who have analyzed these partnerships tend to describe individual "renegade" or "forward-thinking" farmers who, after noticing obvious pesticide pollution and the associated threat of regulatory controls, proactively experimented with alternative pest management practices and sought to coordinate their efforts with other growers in their region.[58] The most successful of these partnerships are typically assisted by sympathetic, environmentally oriented people from other institutions, including staff from small farm organizations, some farm advisers from university Cooperative Extension, leaders from commodity organizations, staff from regulatory agencies like the U.S. EPA and DPR, and independent PCAs. The participants collaboratively develop various practices to reduce pest pressure (such as crop rotations and changed cropping density), conduct experiments, share marketing opportunities, and share knowledge about local ecological relationships between pests, crops, pesticides, and beneficial insects. Hailed as a "community-based approach to environmental protection," such partnerships depend on cooperation, are patterned around "dialogic" social interaction and sharing, and "create synergistic benefits from social learning interactions."[59] Such programs are especially helpful for reducing pest pressures when their participants include a large percentage of farmers from a given region. Also, when designed in a relatively nonexclusive way (e.g., not only for organic farmers or those with small farms), such programs can help to recruit farmers who might otherwise not think of themselves as environmentalists into the project of reducing pesticide use.

Because reducing pesticide use can often increase farmers' workloads and financial costs, many wish to be compensated for taking such extra measures. It is for such reasons that many farmers have been instrumental in developing certification schemes that help consumers "vote with their dollars" by purchasing agricultural products that meet an elevated set of environmental criteria. Farmers participating in such programs arrange to be certified according to the environmental criteria, identify their efforts with a label on the final product, and are typically rewarded with a price premium in the marketplace. Organic agriculture is the most widely recognized of such certification systems; the organic label designates food that has been grown and processed according to a specific set of technical criteria, the heart of which is the avoidance of using synthetic agricultural chemicals.[60] The growth of the organic food and farming sector has helped to reduce the use of the most toxic pesticides and encourage researchers and industry to develop less toxic pest management systems.

Many growers, professional pesticide applicators, and the groups that represent them also participate in voluntary agreements that are designed to reduce the need for environmental regulations. For example, with the assistance of the industry advocacy organization the Alliance for Food and Farming, a group of farmers and chemical manufacturer representatives in California's Kern County recently created the "Spray Safe" program as a nonregulatory way to reduce the number of incidents in which crews of field workers are exposed to pesticides from farmers spraying neighboring fields. The Spray Safe program ("A Farmer to Farmer Commitment to You") promotes voluntary approaches to encouraging careful pesticide applications, keeping workers trained on pesticide safety, and improving communication between farmers about their pesticide applications. The San Joaquin Farm Bureau president argued recently that the program "really helped the agricultural community in Kern County get to the point where they really don't have that kind of drift incident anymore."[61] California DPR awarded a grant to the Alliance for Food and Farming to expand the Spray Safe program to three more counties in California, which will help to reduce pesticide drift incidents by training pesticide applicators to be more careful and better understand pesticide regulations.[62] Similarly, farmers and the organizations that represent them have been active participants in the Ag Futures Alliance, which aims to solve agricultural problems through a "statewide alliance of county-based consensus building roundtables" that was designed to end "polarization and difference."[63]

At the national level, many farming organizations join chemical manufacturers in participating in the National Coalition on Drift Minimization, which aims to address the pesticide drift problem by supplying pesticide applicator training materials and advising pesticide regulators.[64] In addition, the National Agricultural Aviation Association, inspired by "the conscious decision to educate rather than regulate," developed a training program to help reduce aviation fatalities and drift incidents from aerial pesticide applications. The resulting Professional Aerial Application Support System drift mitigation module receives endorsement along with funding from insurance underwriters, pesticide manufacturers, the U.S. EPA, the USDA, and the Federal Aviation Administration.[65]

Structural Barriers to Sustainable Farming
Although all of these efforts to address problems like pesticide drift are commendable, pesticide drift incidents continue. Certainly, pesticide drift and the associated illnesses are often caused by individuals' specific actions, such as farmers who ignore pesticide laws, hired employees who knowingly break rules, or professional applicators who inaccurately judge weather conditions. To some extent, pesticide users' behaviors that exacerbate the problem of pesticide drift can be attributed to a lack of knowledge—hence the proliferation and importance of drift minimization training programs throughout the United States. Yet researchers have found that many participants in such training programs do not actually adopt drift-reducing behaviors, even those participants who have pledged to do so.[66] To understand why this is the case, we must consider the contexts within which farmers and other pesticide users operate—the numerous structural factors that limit their abilities to effectively deal with pesticide drift through voluntary means. As much as individual actors assert their own agency over their lives at work and elsewhere, each of us also is embedded within a constellation of material and cultural structures that shape our actions and opportunities. Pesticide drift is best understood as stemming in part from a wide range of material and cultural structures that compel farmers to apply high rates of chemical pesticides that are frequently prone to drifting into adjoining and nearby neighborhoods as well as other social spaces.

Agriculture tends toward overproduction and thus low commodity prices, since a productive year for one farmer is probably a productive one for most. At the same time, recent structural changes in various nodes of the agrifood industry also put downward pressure on farmers' profits: increasing economic concentration in input sectors has pushed farm input prices up over time, similar concentration within the processing and retail sectors leads to falling farm gate prices, and increasing economic liberalization globally increases the likelihood that California farmers must compete with farmers elsewhere who sell their goods for lower prices.[67] All of these factors compel farmers to produce as much as possible at any cost in order to stay afloat, shifting the risks and burdens of that productivity to other people and spaces in the process.[68]

Growers often claim that it would be economically irrational for a farmer to use more pesticides than necessary ("It's not in our financial interest to do more spraying than we need to").[69] Such statements, however, obscure the material constraints at work here: namely, the fact that losing an entire crop to pests represents a much greater financial loss than do even numerous applications of pesticides. In assessing the barriers to farmers' participation in partnerships that aim to reduce pesticide use, Warner explains that farmers directly benefit from regular pesticide applications: "Growers know that they will not lose their operation if they spray unnecessarily, but know that they may if they do not spray when required."[70] Pesticide applications serve as insurance against crop failure, the threat of which inspires and strengthens the logic of routine, preventive pesticide applications. This is reinforced by the fact that farmers are not generally accountable for any other risks associated with pesticide use and exposure (such as health burdens borne by nearby residents), since it is difficult to definitively link particular pesticide applications with specific illnesses in a context of countless environmental hazards, and because pesticide illnesses are similar to other illnesses and often take years to become manifest. Researchers' interviews with and surveys of farmers show that farmers' pesticide use decisions stem first and foremost from their concerns about maintaining yields.[71] In the words of various farmers interviewed by sociologist Michael Carolan,

I admit I'm torn. On the one hand, I read about the risks associated with pesticide residues on our foods, and that concerns me as a father. But, on the other hand, I can see first hand the effect those chemicals have on my yields. So do I stop using them [chemical inputs] because of what some people think? Or not, and continue to produce at the high end of the yield charts?

It may not be the most PC [politically correct] thing to say, but I need my chemicals. I can't afford to let my yields drop, even by a few bushels per acre.

The first thing that comes to mind when I think of reducing my chemical inputs is lower yields.[72]

Moreover, the "quality standards" that the food industry has developed to differentiate the top-notch produce that will receive the highest market prices generally define quality in terms of the product's visual appearance—not nutrients, product safety, the safety of production practices, or even taste. Such cosmetic standards further compel farmers to increase their pesticide use rates, since even superficial pest damage will usually disqualify a crop from the most lucrative markets. These various circumstances drive not just high pesticide use rates but also compel some farmers to break workplace health and safety rules in order to protect their profits. For example, researchers have found that farmers often violate reentry interval rules (i.e., rules that prohibit people from entering a treated field for a certain time period after pesticides have been applied) and explicitly rationalize that violation as necessary for ensuring the crop's success.[73]

Ultimately, farmers who want to move away from the chemical-intensive model of pest management must do more than reduce their use of chemical pesticides; they must also learn about and adopt a variety of other practices to monitor and manage pests. Little funding is available to assist growers who wish to reduce their pesticide use, though. For example, only 0.3 percent of the most recent Farm Bill's $300 billion is devoted to organic farming. The major grower and trade organizations exacerbate this type of problem by vociferously fighting any federal support for alternative agricultural pest management programs and policies.[74]

Moreover, diversifying one's farm—often a crucial component of reducing pesticide use—is not economically viable where regional economies are structured around single commodities. Reducing pesticide use is also exceptionally difficult where all the local experts and input suppliers are geared toward pesticide-intensive agriculture. Farmers relying on alternative pest management need to draw on experts with different types of knowledge, input suppliers with an unconventional range of products, and fellow growers who encourage and share an interest in agroecological practices. Farming is, after all, a network of social relationships as much as it is a set of chemical and biological ones.

The structural factors that reinforce high rates of pesticide use by farmers and risky practices that contribute to pesticide drift are not just material but cultural, too. In his study of farmers in Iowa, sociologist Michael Bell illustrates that farmers' practices are deeply intertwined with their

social relations and personal identity. Accordingly, the prospect of changing farming practices constitutes a threat not only to their material investments but also to their own sense of self along with all the knowledge,
norms, identity, and social affiliations that entails.[75]

Additionally, there are certain widely held beliefs about what good
farming looks like in the United States: fields that are highly productive as well as free of weeds and insects. Although the clean field ideal
is a socially constructed notion of good farming that flies in the face of
some basic ecological principles, it does serve to reinforce the legitimacy
of chemical-intensive agriculture. Sociological research has shown that
farmers often feel as though they are forever subject to their neighbors'
judgments about the quality of their fields, and adhering to the clean field
ideal is an important part of gaining or maintaining respect within their
communities and families.[76] One PCA explained to me that most farmers
would rather spend time and money making their fields "look like the
cover of a magazine" than "buy their wife a new dishwasher." Commodity
organizations similarly reinforce these particular notions of good farming
that pivot around productivity rather than other goals (just as chemical
manufacturers reify those beliefs through the images in their advertisements). In the words of two farmers interviewed by Carolan,

There's a lot of pressure to keep the corners of the field trimmed and the rows clean.

It's a product of what we in agriculture value. The National Corn Growers Association gives out annual awards for their Corn Yield Contest, and I'm sure the Cattle-
men's Association and the American Soybean Association have similar awards. But
have you ever heard of an award for Best Soil Structure, or something like that?[77]

Researchers who have studied farmers' perceptions of pesticide risks
have found that farmers widely believe that pesticides are safe and controllable, spills during mixing and application are the only significant sources
of pesticide exposure on farms, and pesticide residues do not pose a health
hazard.[78] This widespread casual disregard for the dangers of pesticides
justifies the high rates of pesticide use and risky application practices. Furthermore, such beliefs legitimize inadequate pesticide safety training. Hired
farmworkers receive little pesticide safety education—in some studies, for
instance, less than 20 percent of farmworkers surveyed had received *any*
pesticide training.[79] Even when conducted, pesticide education is usually
inadequate (e.g., conducted in English for Spanish-speaking workers, done
superficially, or done rarely). Many employers avoid giving pesticide training to their workers because they believe that such training unnecessarily
increases fear among the workers to the point that it hinders productivity,

that it is "commonsense" information, and that workers do not want to learn such information.[80]

Actual drift incidents illustrate the consequences of such beliefs. For example, in July 2009, eleven people were sickened in Kern County (and five of those were hospitalized, including two pregnant women) after agricultural pesticides drifted into a nearby field in which workers were harvesting onions. Investigators later found that the pesticide applicator had not received adequate safety training.[81] Indeed, violations of pesticide safety training regulations are frequently identified in investigations of pesticide drift illnesses.[82]

It is not uncommon for farmers to hire certified professional pesticide applicators and other workers to conduct the actual pesticide applications. While these other pesticide users are not subject to the same material structural constraints as farmers are, they have little ability to intervene in pest management decisions in ways that can reduce pesticide drift. In fact, some professional applicators have noted that current regulations grant applicators tremendous discretion and make them legally responsible, and yet they operate under considerable pressure from farmers to spray even in conditions in which drift is likely. DPR's minutes from one of its Drift Minimization Initiative meetings in 2001 with various industry representatives illustrate this dilemma:

> One applicator pointed out (and others agreed) that they would welcome some generic requirements that would prohibit spraying under certain (presently legal) conditions where they know drift is likely to occur, but the grower is demanding they spray anyway. At present, if they want to keep the job, they have to spray.[83]

Research also suggests that certified pesticide applicators underestimate the health risks associated with pesticides—an orientation that legitimizes conducting pesticide applications that have a high potential for drift.[84] Like growers, certified pesticide applicators responsible for pesticide drift incidents and illnesses are typically cited for careless application practices as well as violating pesticide safety training regulations.[85]

Hired workers on farms—mixers, loaders, handlers, and the other workers who actively participate in pesticide applications—are similarly intimately involved with agricultural pesticide use. Researchers have found that hired workers most closely involved with pesticide applications—that is, those who are most likely to be exposed to them—express serious concerns about the risks of pesticides. Researchers have discovered that despite the fact that they have heightened perceptions of risk over the general population, farmworkers feel a relative lack of control over their own exposure to pesticides.[86] Hired farmworkers are relatively unable to intervene in pest management decisions in ways that significantly reduce problems like pesticide drift. The farm labor market is oversaturated, so workers either follow their employers' commands or risk losing their jobs to other, more complacent workers.[87] Accordingly, farmworkers with relatively lower socioeconomic status report feeling a lower sense of control over pesticide exposure.[88]

These power dynamics exacerbate the problem of pesticide drift in various ways. Workers hired to apply pesticides feel unable to critique or question risky practices. Workers follow their supervisors' orders to enter recently sprayed fields before it feels safe to do so.[89] The following statement from a worker who removes the plastic tarps after methyl bromide fumigations highlights why workers keep working through unsafe conditions:

So when the time came to break plastic, you could right away feel the chemical substance. But because we needed the work and we had to do it. And because the chemical was so strong, I think, my nose would start bleeding and my eyes would get very watery and very red all day long.[90]

The fact that hired workers who feel that they are at risk of exposure to pesticide drift are often unable to protect themselves became evident in the Kern County incident described earlier in this section; when the exposed workers reported smelling a strong chemical odor and suddenly exhibited symptoms of pesticide exposure (vomiting, sharp headaches, and burning eyes), their crew boss instructed them to keep working and failed to take them to a physician.[91] A journalist who interviewed the workers later reported that "even after the drift started, the organic farm's supervisor encouraged them to keep bunching onions, telling them to put handkerchiefs over their mouths to block out the smell of the insecticides."[92]

Field workers' relative lack of power does not, in and of itself, cause pesticide drift incidents. Nevertheless, stories like these provide a window into the production and social pressures that farmers and other pesticide users face, how those pressures shape the decisions they make about pest control, and the ways that certain bodies end up bearing the burden of those decisions. Poisoned workers and large-scale pesticide drift incidents are the most visible, visceral, and startling outcomes of a subtle, pervasive problem: industry dynamics and cultural structures that compel farmers and other pesticide users to rely on chemicals as a primary means of squeezing as much value as possible from agriculture at all costs. We can most clearly see these dynamics and some of their immediate costs in cases where farmers and crew leaders instruct groups of workers to keep working even though they are vomiting, losing consciousness, and exhibiting other symptoms of pesticide exposure. These striking cases are thus best understood as the tip of the iceberg—as markers of contaminated landscapes whose human health consequences are certain yet rarely visible, and thus ultimately unquantifiable. As commendable and helpful as industry's initiatives like pesticide applicator training programs and alternative labeling systems are, they are outmatched in many ways by this collection of deeply rooted material and cultural structural supports for pesticide use and regulatory violations.

Industry Self-Representations

Though chemical manufacturers, distributors, farmers, and other actors within the industry play different roles and often work in direct competition with each other, they all profit by keeping pesticides widely available and in high demand in the marketplace. Industry constantly works to legitimize agricultural pesticides—to present them as both necessary and beneficial. In designing their promotional materials and advertisements, industry actors tap into predominant social values to simultaneously promote their sustainability work and obscure their strident defense of existing, highly toxic pesticides. Manufacturers, distributors, and farmers organizations all rely on several common rhetorical themes—environmental sustainability, humanitarianism, and family farm survival—to garner legitimacy, protect their investments, and generate sales.

As public concern about the environmental impacts of pesticides has increased, so too have industry actors' claims proliferated about the environmental sustainability benefits of crop protection products in their advertisements and annual reports as well as on their Web sites.[93] The promotional materials of CropLife America are particularly revealing. Billing itself as the "voice of the industry," and the "largest trade organization for agriculture and pest management" developers, manufacturers, formulators, and distributors, the organization's Web page flashes a seemingly endless loop of environmental sustainability claims at the viewer:

Pesticides enable farmers to produce more crops with less tillage, reducing deforestation and conserving natural resources, while reducing soil erosion. . . . Crop protection products contribute to higher production on good farm land, allowing for return of marginal land to wildlife habitat. . . . Today's farmers use environmentally friendly techniques—such as buffer strips, reduced-till and no-till and precision application of fertilizers and pesticides to prevent run-off. . . . Crop protection products protect water quality and aquatic habitat by reducing soil erosion. . . . Pesticides protect native plants and animals from invasive plants and are used in conservation tillage programs that reduce soil erosion.[94]

Industry representatives emphasize that farmers care about the environment, farming is an inherently environmentally protective land use, the "worst" pesticides are no longer used, and air pollution from agriculture pales in comparison to air pollution from other sources (especially cars, trucks, residential fireplaces, and indoor pollutants). The following statement by the leader of a prominent growers organization vividly portrays industry arguments:

> The farmers have been responsible for air quality for years! We grow plants. What do plants create? If we didn't have these plants, what would this [air] basin be like today? You couldn't live here. It would be a dust bowl. . . . If it wasn't for plants, scrubbing the ozone, it would be even worse! Now the question the public needs to say, do you want agriculture, or do you want Los Angeles? . . . Are they careful about pesticides? They certainly are. They're concerned about the environment. There is no doubt about that. We created a voluntary concept approach to dealing with the air issues. . . . Don't rationalize us as a bunch of polluters! Because the farm is not polluting. . . . [O]ur farmers do really care on what they do with the environment because they live in it.[95]

Assertions like these stand in stark contrast, of course, to the work of environmental historians who have long shown the exceptionally disastrous effects of modern agriculture on the California landscape.[96] These claims are not necessarily false—pesticides can help to reduce other environmental problems, and many farmers certainly do care about the environment—but they can be misleading. They obscure, for example, the regular applications of highly toxic pesticides despite the availability of less toxic alternatives, deny the fact that using pesticides to reduce tillage does not necessarily "protect water quality," hide the structural factors that compel environmentally unsound practices, and neglect to mention industry organizations' vehement rejection of nearly any mandated environmental regulations. Additionally, researchers have found that pest control companies widely use false and misleading statements about pesticides as "safe" in their advertisements.[97] Instead, industry promotional materials paint pesticides as essential not just to environmental protection but also, accordingly, to human health and safety:

> Crop protection products help produce an abundance of affordable, healthy fruits and vegetables which reduce the risk of some cancers and chronic diseases. . . . Pesticides reduce insect-transmitted diseases; protect consumers against dangerous molds; protect pets against fleas and ticks; and control weeds on rights-of-way, thereby contributing to safer driving.[98]

Industry advertisements and other promotional materials also make sweeping humanitarian claims, presenting agricultural pesticides as essential to combating the tremendous problems of hunger and population growth. Industry organizations widely characterize current pesticide products and the rates at which they are used as essential to the production of food, thereby portraying farmers as benevolent philanthropists helping to free millions of people from hunger.[99] Bayer CropScience's Web site features a "population counter" that ticks upward every few seconds, next to headlines that Bayer products are "The second green revolution: Safeguarding food for a growing world population."[100] CropLife America's punchy snippets drive this point home:

Pesticides combat global malnutrition and starvation and help families worldwide afford more fresh fruit and vegetables. . . . With the world population projected to reach nearly eight billion by 2025, pesticides will remain a major contributor in meeting worldwide demand for food and fiber. . . . Pesticides help ensure that consumers have food that is safer, more nutritious, and affordable than ever before.[101]

DuPont's tagline brings the environmental sustainability and humanitarian messages directly together by urging you to "Count On DuPont: Developing the technology and offerings to improve sustainable agriculture for a growing world."[102] Similarly, the CEO of chemical manufacturer Arysta Life Sciences prefaced his defense of his company's new soil fumigant, methyl iodide, by stating, "I'm the proud manager of a company that puts food on every American's plate."[103] Such arguments misleadingly frame hunger as a function of absolute quantities of food produced in the world rather than a function of poverty. In fact, millions of people suffer from chronic hunger in the United States, the land of agricultural plenty.[104] Industry supporters instead tap into contemporary fears as the basis for making misleading connections between current pesticide use patterns, food security, and food safety. One California assemblyperson (and farm owner) defended methyl iodide by saying,

Now I know none of us want to become dependent on foreign entities for our food supply, like we are with our oil supply. But I can tell you that if we are not careful between our water situation and the regulatory burdens that are being placed on agriculture today, that is exactly what is going to happen. And last I checked, we were getting dog food from China and it killed some of our dogs. I do not think we want to go there with our food supply.[105]

In addition to the environmental and humanitarian claims, various industry actors work to legitimize agricultural pesticide use by framing it as essential to the survival of the family farm. Chemical manufacturers and grower organizations alike draw on the pervasive agrarian ideal, framing regulatory restrictions on agricultural pesticides as pushing family farmers out of business. As the vice president of the Western Growers Association stated at a legislative hearing in 2009 on proposed soil fumigant regulations, "The many regulatory mandates and costs associated uniquely with California—and our separate and lengthy process for crop protection registration is one of them—are combining to force small farmers into consolidation with large farms and increasingly motivating farmers to look elsewhere for future operations." There is certainly some truth to this statement. In particular, many environmental regulations pose financial burdens on growers, and these burdens are often felt most acutely by small farmers who are already relatively disadvantaged in the marketplace and operate on lower profit margins. That said, such framings obscure a few important facts: many other factors further marginalize small-scale farmers (including corporate consolidation among input manufacturers, processors, and retailers), chemical manufacturers and distributors stand to lose little by farm structural change, and they invoke the idealized notion in the United States of the family farmer largely to protect their own investments in agricultural chemicals.

Justice and the Industry

Author   So do you think voluntary measures are effective?
Industry leader   I think they're fantastic![106]

Industry actors engage in all these specific material and representational practices to protect their investments. Self-protection and profit maximization certainly do not constitute a theory of justice, nor would many companies' practices be acceptable in any major theory of justice. Still, the agrichemical industry's preferred "solutions" to the problem of pesticide drift interact with a libertarian theory of justice in several consequential ways.

Libertarians believe that the protection of private property rights is the moral basis for a just society, and actors are entitled to freedom from state intervention in deciding whether and how to dispose of their goods and services. Accordingly, the libertarian vision of a just society is that of a grand marketplace unencumbered by state intrusion except for minimal basic protections of private property. Indeed, libertarians argue that state intervention into private affairs is in almost all cases simply unethical, equate taxation to slavery, and find material inequality to be (perhaps) lamentable but ultimately irrelevant to justice. In the libertarian framework, the role of the state is solely to protect "against violence, theft, and fraud, and to the enforcement of contracts."[107] Libertarianism provides moral justification for property owners' increasing frustrations with regulations, taxation, surveillance, and other state interventions that attenuate their property rights and constrain their profits. As David Harvey notes, these frustrations find easy purchase within countries (like the United States) that have long identified as "individualized, utility-maximizing, property-owning democracy."[108]

The agrichemical industry's favored approaches to addressing the problem of pesticide drift directly align with this libertarian vision of a just world and capitalize on its growing popularity. Chemical manufacturers, distributors, and users widely dismiss the need for state intervention in addressing problems like pesticide drift. They insist instead that voluntary and market-based mechanisms can do the trick. Hence, industry actively supports and participates in environmental sustainability indexes (so that investor activity can indicate whether companies are acting in accordance with social preferences, or so the argument goes) and pesticide applicator training programs ("education rather than regulation") at the same time that it devotes massive lobbying expenditures to defeating proposed environmental regulations.[109] Industry is explicit in its rejection of state involvement: regulations are inherently unnecessary, and the market will steer industry practice to provide what the public wants. In the words of a grower organization leader, "Consumer demand tells us not only what commodities to produce but whether to produce them conventionally or organically. These are regulatory forces that cannot be overruled by regulatory fiat."[110]

Such programs and practices are quite limited in a few ways, though. Voluntary and market-based programs are inherently spatially and temporally uneven—attending to the problem (however [in]effectively) in some places but not others, and ebbing and flowing over time as interest and resources permit. Moreover, the specific practices pursued by chemical and agricultural industry actors fail to address many of the key factors that make pesticide drift such a serious problem. Training applicators to more carefully adhere to pesticide regulations, for example, can surely help to reduce large-scale pesticide drift incidents, yet it does little to reduce the overall ambient pesticide load to which agricultural community residents are exposed on a daily basis, redress the structural forces that compel current pesticide use practices, or ameliorate the social and economic forces that compromise compliance with health and safety regulations. Because of these limitations, voluntary and market-based practices on their own do not constitute adequate mechanisms for addressing pesticide drift, or many other environmental problems.

Libertarianism has also been criticized as inherently undemocratic since it overlooks the ways in which social inequalities inhibit the ability of the poor and otherwise-marginalized groups from participating on equal footing in the marketplace, public life, or the policy arena. Philosophers widely fault libertarians' propensity to ignore groups entirely; Young hence criticized libertarianism's "possessively individualist social ontology."[111] Moreover, libertarians dismiss the immensely unjust circumstances in which most wealth and property in the modern world was originally acquired, and provide little guidance on how such injustices in acquisition could ever be even remotely rectified, which suggests that private property is simply an indefensible basis for justice. In general, despite their increasing popularity in the environmental realm, libertarian ideas and rhetoric do not help us to effectively control many environmental problems.

Additionally, a libertarian notion of socially just environmental problem solving does not provide any tools for addressing situations in which protecting one actor's private property directly infringes on another's, and where no market exists for those actors to fairly reach a compromise—as occurs with pesticide applications and nearby bodies when pesticides drift. Contexts marked by extraordinary inequality end up obscuring this contradiction, in turn giving libertarianism more traction. For instance, in the case of pesticide drift, infringements on the private property rights of industry (i.e., pesticide regulations) are apparent and bound to be contested by powerful actors. In contrast, infringements on the private property rights of agricultural community residents (i.e., chemical trespass into their bodies) are relatively invisible, nearly impossible to definitively determine, and harm actors with relatively little social power. A libertarian approach to solving problems like pesticide drift therefore effectively defends and protects the private property rights of the most advantaged while infringements on the rights of those with relatively little power remain hidden from view. In turn, a libertarian approach both *appears* to solve the problem and ends up reinforcing the highly unequal status quo.

It is for such reasons that political philosophers have derided libertarian notions of justice and the policies they have (at least partially) inspired for resting on overly simplified ideas of private property along with unfounded assumptions about the ability or propensity of property owners to protect against environmental harms. Such notions grossly underestimate the extent to which environmental problems are externalized to the public (or other species, other places, or future generations) and overestimate the ability of economic competition to ameliorate such problems. Political philosophers thus have widely characterized libertarian conceptions as "intellectually bankrupt," "incoherent," and "wrong."[112]

Although industry's so-called sustainability practices are limited in their ability to meaningfully address problems like pesticide drift and align with a highly problematic theory of justice, libertarian ideals and practices continue to proliferate and gain legitimacy in nearly every other sphere of social and political life. As we will see in the next chapter, market-based and voluntary programs are rising in prominence in the regulatory arena at the same time that environmental regulatory protections are being slowly eviscerated. These trends are the two sides of the libertarian coin. Before I turn to the regulatory arena, however, it is worth looking briefly at the work of pesticide drift activists to directly confront industry's role in the problem of pesticide drift.

Pesticide Drift Activism and Corporate Accountability

Rachel Carson is often credited with sparking the modern environmental movement in the U.S. and elsewhere through raising awareness about DDT and other chemicals in *Silent Spring*. But many environmental movements may have missed an essential message when she wrote of "an era dominated by industry, in which the right to make money, at whatever cost to others, is seldom challenged" (Carson 1962, 43). That is, tackling the threat to public and environmental welfare is not just a matter of curbing particular corporate harms, or even creating and promoting sustainable alternatives. Ultimately, the structure of corporate rights and power must be addressed.

—Skip Spitzer, "Industrial Agriculture and Corporate Power"

Pesticide drift activists widely promote organic agriculture and other forms of less toxic pest management as the means through which farmers can directly reduce pesticide drift. In so doing, they dispute the notion that current pesticide use is an essential part of food and farming. Some pesticide drift activists also participate in various industry-sponsored, education-based efforts to address pesticide drift (such as the Spray Safe program). Nevertheless, pesticide drift activists generally collaborate very little with the agrichemical industry, instead vociferously disputing industry's calls for greater insulation from government restrictions. Pesticide drift activists' attention to the agricultural industry critically confronts the ineffectiveness of the industry's voluntary efforts to address pesticide drift. For example, one activist leader characterized the Spray Safe program as "total greenwashing. . . . It encourages growers to communicate with each other and use common sense, but it doesn't do anything besides recommending that people follow the law. Also, it is self-selecting. It attracts people who already are relatively safe."

Through their campaign materials, at public regulatory hearings, and in opinion pieces to the media, politically active pesticide drift victims recount their own experiences of pesticide drift along with the lingering illnesses, debt, and fear—stories that serve as a continual implicit reminder that industry actors' good intentions and claims about sustainability do not protect people against pesticide drift. On its Web site and in its campaign materials, PAN points out that the industry actors that tout their own sustainability practices are also complicit in and responsible for a wide array of scandalous environmental damages.[113] PAN's corporate profiles for the major agrichemical manufacturer Dow, for instance, contrasts the corporation's sustainability claims with a laundry list of irresponsible practices, including suppressing its own research from the 1960s that showed that its pesticide dibromochloropropane (DBCP) caused sterility in laboratory rats; continuing to manufacture and export DBCP after it was banned in the United States in 1979 due to its irrefutable contribution to human sterility; failing in many ways to assist the tens of thousands of victims of the chemical disaster at a factory of its subsidiary (Union Carbide) in Bhopal, India (instead, suing victims who peacefully protested outside its corporate headquarters); dumping hazardous wastes; and falsely advertising some of its most toxic products as safe. Similarly, major NGOs involved in pesticide drift activism point out that industry's investments into genetically engineered crops actually help to reinforce (rather than supplant) the centrality of pesticides in agricultural pest management.[114] PAN's work also explicitly highlights the fact that the same corporations that produce and sell carcinogenic agrichemicals also produce cancer-treating pharmaceuticals.[115]

In publicizing such contradictions and malfeasance, the major NGOs in the United States working on pesticide drift have served as key members of the International Campaign for Justice in Bhopal, the Dow Accountability Network, and the Agribusiness Action Initiatives.[116] One example of this type of effort is the recent campaign by PAN and other organizations to direct the U.S. Department of Justice to investigate the near-monopoly power held by agrichemical giant Monsanto.[117]

At a local scale within the United States, pesticide drift activists write letters to the editors of local newspapers in the wake of specific pesticide drift incidents, demanding that pesticide applicators and other complicit actors be held liable for their regulatory violations. Pesticide drift activists also provide moral support to people who suspect they have been exposed to pesticide drift, and help supply legal support to victims who decide to sue the parties responsible for their illnesses and medical bills. Activists note that unchecked corporate power makes pesticide drift incidents just a small cost of doing business for many of the industry actors involved; even while lauding fines and court settlements following pesticide drift incidents, activists point out that the sums generally represent a mere drop in the bucket for the responsible parties relative to their revenues. Activists underscore the inequalities and contradictions inherent in industry actors' propensity to both minimize concerns about public health and express outrage about crop damage from drifting agricultural chemicals. For example, in reaction to a court verdict that awarded $7.5 million to a grape grower whose vineyards were, according to the local paper, "severely damaged" and "chemically burned," one activist leader commented,

When people are poisoned by pesticides, nobody believes that they're even sick, let alone saying that they were "chemically burned," or mentioning that it's very difficult to get pesticide drift investigations conducted when community members are poisoned. Even when investigations are conducted and find violations, fines are never issued in the vast majority of cases and are generally for no more than a few hundred dollars—certainly not $7.5 million![118]

Activists' efforts thus undermine industry claims to responsibility and sustainability, and the countless lawsuits that have been brought against chemical manufacturers and applicators further substantiate these convictions.[119] Using these tactics, pesticide drift activists work to show that industry itself cannot be relied on to control the pesticide drift problem through its voluntary and educational programs. Pesticide drift activists contest the way that agricultural community residents have been abandoned and sacrificed by the push to rely on market signals as well as friendly educational campaigns, however well intentioned and helpful those efforts are. By illustrating the limits of a libertarian approach to environmental problem solving, pesticide drift activism weakens the increasingly predominant conception of justice bolstered by industry's approaches to solving environmental problems. Consequently, pesticide drift activists argue that pesticide drift is sufficiently critical that only the state is in a position to effectively address it, and they direct their anger and energy toward regulatory reform. In the next chapter, I follow the activists' gaze to the regulatory arena, describing the regulatory context and activists' efforts to reconfigure it. As will become clear, the activists' critique of the state's approach to addressing pesticide drift contains not just a condemnation of libertarian-inspired solutions but also a fundamentally different vision of what a socially just approach to environmental problem solving looks like.

Conclusion

Chemical manufacturers, distributors, applicators, and farmers all participate in a number of practices to help address the problem of pesticide drift. Yet as we have seen, such practices have significant limitations, those who employ them face numerous constraints in doing so, and the practices are overshadowed by numerous structural factors that undergird and reinforce chemical-intensive crop protection. Engaging a variety of narrative tropes, however, enables industry to simultaneously promote its sustainability practices and obscure its strident defense of decidedly less-than-sustainable investments. The agrichemical industry's approach to addressing pesticide drift also gains legitimacy by aligning with the increasingly predominant libertarian notion of justice. As this perspective on justice requires minimizing state infringements on private property rights, voluntary and market-based programs emerge as the only appropriate solutions in spite of their practical limitations. That industry embraces such approaches is not at all surprising. What is surprising is that this ideological trend, with all its abandonments and sacrifices, also pervades the regulatory arena and activist circles. The next two chapters will show how this has occurred, how these ideological trends both exacerbate and obscure pesticide drift, and how pesticide drift activists marshal a different notion of justice to more effectively combat pesticide drift.

4

The Environmental Regulatory State

Before Love Canal, I also needed a 95 percent certainty before I was convinced of a result.[1] But seeing this rigorously applied in a situation where the consequences of an error meant that pregnancies were resulting in miscarriages, stillbirths, and children with medical problems, I realized I was making a value judgment . . . whether to make errors on the side of protecting human health or on the side of conserving state resources.

—Beverly Paigen, biologist with New York State Department of Health, 1982

As the preceding chapter showed, industries are not able to effectively control problems like pesticide drift on their own. State institutions are needed to serve that function. In this chapter, I discuss the ways that government agencies have intervened into agricultural pesticide use. California and federal pesticide regulatory agencies are some of the most advanced in the world. Yet the case of pesticide drift and illness presented in chapter 2 raises questions about how well regulatory agencies control pesticide drift. This chapter takes on this task. Through describing the history of pesticide regulation in the United States, I identify the material, cultural, and discursive factors that explain regulatory agencies' failures to adequately address the pesticide drift problem. As will become clear, this foray into the world of pesticide regulation is a story about agencies that are captured by industry and hobbled by internal dysfunctions, but at the same time, are pursuing a set of environmental protections that align with a particular vision of justice.[2] I close the chapter with a section on pesticide drift activists' efforts to reform U.S. and California pesticide regulation. First, however, to understand pesticide regulation today, we need to start at its inception over one hundred years ago.

The Early Years of Pesticide Regulation

As environmental historians have shown, members of the public have periodically voiced concerns about pesticide residues in food, and agricultural community residents occasionally reported illnesses that they attributed to nearby pesticide use. Though these reports sometimes inspired state investigations of agricultural pesticide use practices, the state did not begin regulating pesticides to curb their environmental and public health impacts until well into the latter half of the twentieth century. In fact, pesticide regulation through 1970 was limited to protecting industry rather than public health or the environment. The nation's first pesticide laws were crafted in response to growers' complaints about pesticides that were ineffective and others that damaged crops.[3] Consequently, throughout the first half of the twentieth century, Californian and U.S. authorities regulated agricultural pesticide use by requiring pesticide manufacturers to register their products and promise that they were safe and effective.

To some extent, regulatory agencies' apparent disregard for the environment and public health along with their protection of industry interests were rooted in the structure of the regulatory institutions. At the state and federal levels, pesticide regulatory responsibilities rested in the Departments of Agriculture, whose primary responsibilities were to promote and protect the agricultural industry. Although later regulations required the USDA to take input from the Public Health Service into consideration when regulating pesticides, the USDA itself maintained the sole decision-making authority with regard to agricultural pesticides at the federal level. The USDA's conflict of interest—regulating and promoting the same industry—led regulators to rely on educational tactics, rather than direct punishment, for ensuring compliance with the minimal regulations.

Historians have also shown that regulatory failure stemmed directly from malfeasance and corruption within regulatory institutions. Pete Daniels provides detailed evidence of the ways in which USDA regulators actively and publicly championed chemical solutions to agricultural problems, routinely ignored input from the Public Health Service and the Food and Drug Administration, and refused to cooperate or share information with any of those other agencies with which it was required to do so.[4] The fact that state agencies protected industry interests so well stemmed to some extent from a long-standing revolving door between the two social worlds: industry representatives took jobs within regulatory agencies, and state scientists went on to start their own pesticide manufacturing companies.[5]

Yet the story of early pesticide regulatory priorities is not simply one of corrupt bureaucrats and interest group politics. Trends in pesticide policy and regulatory practice must be understood as shaped by the predominant social beliefs and issues of the day. In general, environmental historians have shown that pollution had long been perceived as central to progress, economic growth, and prosperity.[6] Bosso emphasizes that one of the top issues on the political agenda throughout the first half of the twentieth century was "the farm problem"—figuring out how to maintain farm productivity and raise farmers' incomes at the same time.[7] Farmers, legislators, regulators, and the general public widely adhered to the "pesticide paradigm," viewing chemical pesticides as a crucial and liberating way to maintain farmer incomes, produce food for a growing population, and boost the country's standing in the global economy. In these ways, the pesticide paradigm functions ideologically, since this set of beliefs reinforces existing institutions and practices. Regulatory agencies' failure to address the environmental or human health consequences of agricultural pesticides therefore has had as much to do with widespread social beliefs as with misconduct or malfeasance.

Regulatory agencies' protections of industry interests were supported by the legislature. Throughout the twentieth century, Congress was highly sensitive to the needs of farmers, who were politically well organized by the Farm Bureau, constituted a powerful lobbying force in Washington, DC, and (in the early half of the century) represented a large percentage of the U.S. population.[8] Environmental groups that did exist were preoccupied with wilderness preservation—not with polluted, working landscapes—and thus had much less influence over legislative and regulatory practice when it came to agriculture.

Of course, the state's relationship with pesticides was not limited to its industry-protective pesticide regulations. As noted in chapter 3, the state actively helped to develop and promote chemical technologies for both warfare and peacetime purposes. State institutions also have further solidified the industrial model of agricultural production that exacerbates pest problems and hence pesticide use in many ways—through other agricultural policies such as subsidies and tax breaks for monocrop production, food policies, public university research priorities, university extension practices, and intellectual property rights law.[9]

Regulation Reborn

By the 1960s, the chemical foundation of U.S. agriculture had begun to crack. Many widely used pesticides had lost their effectiveness, as the target pests developed resistance or were replaced by other "resurgent" insects that similarly damaged crops. Additionally, several major food safety crises associated with insecticides prompted the public to demand greater regulatory attention to the health effects of chemical pesticides.[10] Pesticides' impacts on wildlife had become more apparent as well. The publication of Carson's *Silent Spring* in 1962, as noted earlier, played a major role in the elevation of pesticides on the public and political agendas, as her writing illustrated the ecological devastation of chemical pesticides in terms that were accessible to the public and policymakers. Carson's scientific research and eloquence further strengthened her testimony before Congress in the 1960s on the pesticide paradigm's drawbacks, the USDA's conflicts of interest, and its associated failure to regulate pesticides.[11]

It was within this context in the early 1970s that Congress passed a host of environmental laws and created the U.S. EPA to more effectively regulate environmental issues. Pesticide regulatory responsibilities were shifted from the USDA to the EPA's Office of Pesticide Programs, which was authorized to register and regulate pesticides under the revision of the Federal Insecticide, Fungicide, and Rodenticide Act (FIFRA) in 1972.[12] Specifically, FIFRA requires the U.S. EPA to determine whether the costs of any given pesticide are reasonable in light of the expected benefits and restrict the use of pesticides accordingly.[13]

Within the United States, states are required to uphold federal pesticide laws and regulations and are authorized to implement additional, more stringent environmental protections. Most states' pesticide regulatory functions are fairly limited in size and are located within the local department of agriculture, thereby institutionally situated as part of industry promotion and separate from environmental protection. Only a few U.S. states have moved those responsibilities from agricultural to environmental departments.[14]

The U.S. EPA's pesticide program and its state-level counterparts generally approach the task of regulating pesticides and other chemicals through managing (rather than preventing) pollution, and they do so by assessing the risks posed by pesticides, designing risk management strategies to minimize those risks, and enforcing the associated regulations. Regulatory agency risk assessment processes for new pesticide registrations (and occasionally, continuing registrations for pesticides already in use) generally reflect the risk assessment principles presented in chapter 2. Regulatory scientists draw on toxicological research to determine the chemical's potency and design a set of health benchmarks. They use those health benchmarks to evaluate existing exposure data (such as air-monitoring study results and modeling estimates). This analysis is conducted by federal and some state agencies to determine whether the pesticide poses risks that need to be mitigated with use restrictions designed to keep the exposure levels below the health benchmarks. Risk assessment is a relatively new institution within environmental regulation. It was formalized in the 1980s and implemented to standardize regulatory scientific protocol as well as bring greater accountability and rationality to regulatory decisions.[15] The U.S. EPA and California DPR emphasize that their health and environmental protection standards have become stricter over time and that they periodically reevaluate registered pesticides according to their latest standards and updated scientific research.

The U.S. EPA uses risk assessment to design pesticide use regulations that theoretically will keep the ambient concentration of each pesticide below a level of public health concern. Those regulations are then specified on the product label, and the label language constitutes the federal law for each product.[16] Risk mitigation strategies for any given pesticide can include a variety of mandated or recommended restrictions on the pesticide's use (including amount, timing, approved uses/crops, frequency, application method, and/or buffer zones around sensitive sites), reentry intervals and personal protective equipment (for worker health and safety), and clarified use instructions. State and federal regulatory agencies typically consult with industry throughout the process of designing mitigation strategies, and state and federal agencies often consult with each other to streamline the process. Pesticide regulations and guidelines are designed so that if they are followed by the applicator, the product will not cause adverse effects for human health or the environment.

The states hold the primary responsibility for enforcing federal pesticide laws. In California, county agriculture commissioners (CACs) are the primary enforcers of pesticide regulations, holding broad power to manage both pests and pesticides. In addition to their other responsibilities that do not pertain directly to pesticide use, CACs and their staff oversee the use of specially regulated pesticides, periodically inspect pesticide applications, investigate pesticide drift and illness reports, administer penalties and other enforcement actions, help process the state's pesticide use reporting system data, administer educational programs to growers and other pesticide applicators, and help to control invasive pests.[17] This devolution of pesticide regulatory authority stems from broader Progressive Era "direct democracy" reforms that took power away from centralized agencies and legislative processes, and instead favored decision making through local county agricultural boards, commodity commissions territorialized to particular production areas, and water boards covering single watersheds. As a result of such reforms, water, land use, and agricultural regulation have rested largely in local hands for over a century in California.[18]

State and federal agencies stress the health-protective measures that they take. For example, the U.S. EPA claims that it "ensures that each registered pesticide continues to meet the highest standards of safety to protect human health and the environment."[19] Similarly, DPR representatives point out that current regulatory structure and practice are both sufficient and exemplary. DPR proclaims that it has "broad authority over the registration, sale, and use of pesticides in California," with "wide latitude to regulate" such that "the Department can, with sufficient reason, demand that all use of a chemical cease immediately."[20] Though pesticide regulatory agency staff members acknowledge that more could be done to better address the problem of pesticide drift, regulatory agency representatives' official stance is that the issue is under control. The following statement from the then-director of DPR in a television documentary aired in 2004 exemplifies this "party line" to which agency representatives dutifully adhere in public comments:

The California pesticide regulatory program is the best in the world, and we have a law that says you can't use a pesticide to cause an environmental effect or a human health problem. Period. So that's the law. And I'm not sure you could make it stronger than that.[21]

State and federal pesticide regulatory agencies have undeniably come a long way in the past century. Some of the agencies' staunchest critics remarked in interviews with me that DPR and the U.S. EPA have improved considerably in the past ten years in their regulation of pesticides. The U.S. EPA has tremendous capacity to regulate, as its Office of Pesticide Programs employs 920 full-time equivalent staff members and, in 2008, operated with an annual budget of $139 million.[22] California DPR is widely regarded as the nation's largest state-level pesticide regulatory program; as of 2009, it has approximately 385 staff members (who collaborate with over 300 employees of CAC offices), an annual budget of $73 million, and a progressive record of environmental protections that often drive federal regulations.[23] New legislation has strengthened the pesticide regulatory agencies' abilities to protect public health and the environment.[24] These agencies devote extraordinary resources to assessing risks, designing mitigation strategies when the risks exceed the health benchmarks, and enforcing pesticide laws and regulations.

The U.S. EPA has made numerous accomplishments in the area of risk assessment in recent years. Notably, in 2009, the U.S. EPA completed its "reregistration review" of all pesticides registered before 1984 (as mandated by the amendment to FIFRA in 1988), and the agency has started the process of reevaluating the registration of all pesticides every fifteen years. As part of the reregistration review, the U.S. EPA extensively evaluated the risks posed by soil fumigants. Additionally, the Food Quality Protection Act (FQPA) of 1996 required the U.S. EPA to revise its risk assessment process for all pesticides to more fully account for the cumulative impact of human exposure to all pesticides that share a common mechanism of toxicity; consider the aggregate risks from a given pesticide as it is used in agricultural, commercial, and/or residential settings; include extra protections for children (whose bodies are more sensitive to exposure); require manufacturers to supply data on their products' propensities to disrupt the endocrine system; and use the revised risk assessments to set new standards for pesticide residues in food.[25] In 2009, the U.S. EPA announced its plan to expand these revisions to all other pesticides' risk assessments (i.e., to those that do not pose risks in the form of residues on food).[26] To facilitate the transition to less toxic forms of agricultural pest management, both California and U.S. pesticide regulatory agencies prioritize and streamline the registration process for reduced-risk pesticides.

U.S. EPA investigations have led to the cancellation of many organochlorine pesticides—all of which are highly persistent, prone to accumulating in animal and human bodies, concentrated throughout the food chain, capable of long-range transport, highly acutely toxic, and associated with one or more chronic health problems. Notably, the U.S. EPA recently canceled all uses of endosulfan, an organochlorine pesticide that had been widely shown by research to disrupt the human endocrine system plus contribute to autism and reproductive disorders, was already banned in over sixty countries, had been banned from use in homes and gardens in the United States since 2000, and had been shown to drift off-site.[27] In response to the FQPA, the U.S. EPA has compelled registrants to voluntarily cancel some uses of many organophosphates in order to reduce people's exposures to pesticide residues on food. Moreover, after reviewing the major soil fumigants as a group, the EPA issued new use restrictions in 2009 to curb fumigants' public health impacts. The U.S. EPA has also restricted the use of many pesticides by classifying them as "restricted use products," which can only be applied by licensed pesticide applicators. The California DPR has added additional pesticides to this list (using the term "restricted materials"), and requires that growers wishing to apply restricted materials must first acquire a permit from their local agricultural commissioner each year and register their intent to use any restricted materials with the local agricultural commissioner at least twenty-four hours prior to the actual application.[28]

The agencies' work extends beyond the risk assessment and enforcement processes. DPR has improved its enforcement practices by mandating new enforcement guidelines (to help standardize the actions that CACs take in response to regulatory violations).[29] To improve pesticide illness reporting, DPR has created and distributed community guides in English and Spanish that instruct the public about reporting pesticide drift and illnesses, increased opportunities for public comment on proposed regulatory decisions, and improved coordination between California's Poison Control System and DPR's PISP database.[30] California DPR collects a tremendous amount of data on pesticide use, illness, and regulatory violations, and it has made those data sets increasingly publicly available through its Web site in recent years. Through its Pest Management Alliance Grants program, DPR has distributed ten million dollars since 1998 to over two hundred pesticide reduction projects organized by industry organizations and community groups.[31] Furthermore, as part of California's new EJ mandate, DPR is working to integrate many EJ principles into the department's work and recently conducted a large-scale, participatory, multistakeholder air-monitoring project in the Central Valley. The U.S. EPA also has launched an EJ initiative that includes increasing local-level public participation opportunities and evaluating the fairness of federal environmental regulatory protections.

Regulation in Retreat

The massive pesticide regulatory apparatuses in California and the U.S. EPA present considerable potential for the health-protective regulation of agricultural pesticides. Yet despite their mandates, resources, and accomplishments, California and U.S. pesticide regulatory agencies have failed in numerous ways to protect public health from the adverse effects of agricultural pesticide use. Overall, the U.S. EPA has been notoriously slow at implementing its mandates. In its first fifteen years (i.e., 1970–1985), the agency completed full investigations of only a handful of pesticides. The reregistration process mandated by the FIFRA reforms of 1972 was not completed until 2008, with few of those evaluations conducted prior to the 1990s. For each chemical, the EPA takes *years* to collect data, conduct risk assessments, design risk mitigation plans, submit them for public comment and peer review, and prepare final regulatory decisions. Pesticide registrants (manufacturers) often take several years to supply missing data. During this time, all previously registered pesticides continue to be used without a reliable assessment of their impacts on human health or the environment.

Though the U.S. EPA eventually canceled registrations for DDT and many other organochlorines, it only did so after insect resistance to the products had effectively diminished demand and use, or after many years of relentless public pressure. Additionally, some organochlorines remain in use. Despite the fact that the U.S. EPA has increased its attention to the highly toxic organophosphates and carbamates over the past ten years, the agency's evaluation and regulation of these classes of pesticides have been vociferously criticized by its own scientists, its Scientific Advisory Panel, the Office of Inspector General, the National Academy of Sciences, and other scholars.[32] Although the U.S. EPA banned all residential uses of chlorpyrifos and diazinon in 2001, it did not place any restrictions on their use in agriculture—despite the fact that chlorpyrifos in particular is frequently implicated in pesticide drift incidents and has been shown in monitoring studies to drift off-site at levels that exceed health benchmarks. Malathion and chlorpyrifos rank as the most widely used agricultural insecticides in the nation.[33]

In California, soil fumigants continue to be used in greater absolute quantities than any other group of conventional pesticides.[34] The overall use of soil fumigants in California remains steady, as shown in figure 2.3 in chapter 2. Despite the fact that methyl bromide is being slowly phased out in accordance with the Montreal Protocol on Substances That Deplete the Ozone Layer because of its damaging effects on the stratospheric ozone layer, the total amount used has remained high since 2001 because the federal government continues to grant "critical use exemptions" to commodity groups (notably, those representing the strawberry industry) that claim to have no other viable alternative. Moreover, California growers have been replacing methyl bromide with the other existing soil fumigants, all of which pose extraordinary risks to human health. California PUR data show that the use of 1,3-dichloropropene has quadrupled since 1996; metam sodium is used in greater quantities than any other pesticide except for low-toxicity horticultural oil and sulfur; metam potassium's use increased by 46 percent between 2007 and 2008; and the use of chloropicrin has doubled since 1996. Under extraordinary critique from distinguished scientists and environmental advocates, in 2008 the U.S. EPA registered a new soil fumigant, methyl iodide, even though it is widely recognized as extraordinarily toxic and prone to drift.[35] Such trends suggest that regulatory agencies are much more capable of ushering new products to the market than of restricting the use of pesticides that are known to be exceptionally dangerous—even those for which viable and less toxic alternatives exist.

Though the U.S. EPA's recent review of the major soil fumigants culminated in a set of restricted use regulations, those have been highly criticized. For example, the new regulations continue to allow sprinkler applications of some fumigant pesticides—a dangerously sloppy and widely used application method through which fumigants quickly turn into vapor (even before hitting the soil) and drift off-site, and one that is commonly implicated in pesticide drift events. Surprisingly, in May 2009, the Obama administration EPA diluted the new protections first announced by the EPA under the Bush administration. Although the U.S. EPA's proposed soil fumigant regulations included some buffer (no-spray) zones to protect the health of people living and working near fumigant applications, for instance, the EPA subsequently reduced the size of those buffer zones and stated that they may now lawfully include farmworker housing, farmworker children's play areas, and streets that do not have sidewalks.[36]

Regulatory agencies' failures to meaningfully protect public health from agricultural pesticide use have not been limited to organochlorines, organophosphates, carbamates, and soil fumigants. Many other pesticides that are widely used are subject to only minimal restrictions and pose significant threats to human health. Atrazine, for example, has long been (and continues to be) one of the top most widely used pesticides in the United States despite the fact that it is a major groundwater contaminant, has been shown to disrupt the human endocrine system, has been linked to cancer and birth defects, and is banned in the European Union.[37]

The U.S. EPA has little ability to systematically evaluate the damage inflicted by pesticides already in use, as there are no national monitoring programs for agricultural pesticide use patterns, drift, or illnesses. In 1993 and again in 2001, the U.S. Government Accountability Office (GAO) stridently criticized the lack of a national pesticide illness surveillance program.[38] As Linda Nash has argued, the failure to monitor pesticides effectively dismisses and obfuscates their public health consequences, since monitoring in and of itself signals that a problem exists.[39]

California's TAC act of 1984 is largely responsible for the fact that the California DPR works with other state agencies to monitor ambient air for pesticides. In cases where this monitoring data and recent toxicological research suggest that the pesticide in question may "cause or contribute to increases in serious illness or death, or that may pose a present or potential hazard to human health," DPR may designate the pesticide as a TAC.[40] Though the TAC act serves the important function of requiring ambient air monitoring, the TAC designation has no mandatory regulatory consequences. Moreover, DPR has conducted a full TAC review for only eight pesticides in the past twenty-five years.[41] One particularly striking case is the fumigant metam sodium; although DPR designated it as a TAC in 2002 and although it has been widely implicated in pesticide drift incidents, the department did not propose any mitigation measures until 2009.

DPR's own data show that existing regulations are simply not adequate for protecting public health. In nearly half the reported agricultural pesticide illnesses in DPR's PISP database from 2003 to 2007, investigators were unable to identify *any* regulatory violations.[42] The existing regulations' inadequacy is illustrated in the following description, published in formal regulatory agency documents, of a pesticide drift event in Monterey County in 2007 for which no violations were ever determined:

Vapor escaped from a field fumigated with a mixture of 41.5% chloropicrin and 57% methyl bromide. The application had gone smoothly and had been monitored by a CAC employee who noted no deficiencies. When nearby residents reported eye and respiratory irritation, CAC staff canvassed the affected neighborhood and identified 31 people probably or possibly affected, including two of the investigators. *The investigation identified no cause for the problem beyond the fact of performing a fumigation near a residential area.*[43]

DPR concludes that their violation data summary "indicates the importance of continuing compliance efforts to further reduce pesticide-related illnesses and injuries"—in other words, that its data point to the significance of getting pesticide users to comply with existing regulations.[44] It seems equally prudent, if not more so, to ask why pesticide illnesses occurred in the absence of any regulatory violations.

My analysis of the "violations" identified in DPR's illness data from 2007 further indicates that existing regulations simply cannot protect public health. That year, half of the illness cases were not attributed to any pesticide violation (consistent with the broader trend noted above), and the only violation identified in one-quarter of the remaining illness cases was the simple fact that drift occurred.[45] Many other illness cases that year also were characterized as a violation of California's "general standards of care"—an exceptionally vague regulation that stipulates that applicators must use "due care" in applying agricultural pesticides.[46] In other words, even the determination of violations and fines does not indicate that pesticide applicators deviated from standard, legal application procedures. For example, in 2007, a Kern County metam sodium application drifted off-site and made at least ten people ill (with symptoms lasting for a week for several of the victims). The agriculture commissioner investigators "cited the applicators for exposing the nearby workers, *but found no shortcomings in their procedures.*"[47] The applicators were fined nine thousand dollars for violating the product label, since it says, "Do not apply this product in a way that it will contact workers or other persons, either directly or through drift."

In sum, pesticide regulatory agencies continue to allow many highly controversial pesticides to be used without adequate regulatory controls, thereby contributing to drift and countless illnesses. These regulatory failures and their human health consequences stem from numerous interacting factors, as I explain below.

Legislative Mandate and Regulatory Tools

Agricultural pesticides in general are difficult to regulate for many of the same material reasons that pesticides drift and cause illnesses in the first place: they are sold in tens of thousands of formulations and used in countless combinations; applied millions of times per year across millions of acres by countless individuals; used on many soil types in varying climatic conditions; employed to combat thousands of different insects, diseases, and weeds; applied with a wide variety of drift-prone methods; and are susceptible to transport through wind, fog, rain, and ever-variable weather patterns. The consequences of pesticide use are often spatially and temporally displaced as well—the effects on human or other organisms show up far from the site of application, or do not manifest for many years—so no reliable indicators of pesticide drift or exposure exist. These are the material conditions that pesticide regulators face.

The major pesticide legislation of the early 1970s charged the U.S. EPA with two new, tremendous tasks: evaluating tens of thousands of pesticides that were already in full use, but whose safety had simply never been assessed, and designing regulations for problematic pesticides so that the costs of the regulation would be reasonable in light of the benefits. However sensible and ambitious, these mandates and the tools with which regulators worked to fulfill them systematically privileged the status quo—the nearly unbridled use of highly toxic and drift-prone pesticides.

To some extent, the U.S. EPA's actions on pesticide regulation were stalled throughout the 1970s by a lack of clear direction and internal conflict.[48] Although the FIFRA amendments of 1972 charged the Office of Pesticide Programs with ensuring that pesticides already registered for use fulfill two competing criteria—that they be both safe and effective—the legislation supplied no instructions on how to implement these two mandates or how to set priorities between them. Internal disarray within the U.S. EPA combined with conflict over the agency's priorities further thwarted the early work of the agency and effectively stalled its implementation of FIFRA. All the while, FIFRA's structure allowed for pesticides ostensibly under review to continue to be registered and used until the EPA could make a strong enough case otherwise.

Paralysis in the process of evaluating old chemicals also stems from the U.S. EPA's mandated responsibility to balance the costs and benefits of any proposed regulation. The benefits of pesticides already in use were clearly evident—widely used pesticides had helped farmers control pests and boost agricultural productivity, especially since new products were continually introduced as insects developed resistance to old ones. Because the economic utility of pesticides to a powerful political constituency—the farming community—was well established, restricting the use of any pesticide would require a strong scientific case that its costs outweighed its benefits. In fact, despite having evaluated only a handful of chemicals in the first few years of its existence, the EPA was ordered by Congress in 1975 to pay even greater attention to the benefits of agricultural chemicals.[49] Subsequent administrations have continually reinforced this mandate. Notably, Ronald Reagan issued Executive Order 12291 in 1981, greatly expanding this charge and establishing the Office of Management and Budget to enforce it.

Pesticides' benefits were apparent, easily calculated, and staunchly defended. The fact that the EPA has historically only compared a pesticide with its major chemical alternative(s) (i.e., not less toxic systems of pest management) artificially inflates the benefits of a given chemical.[50] In contrast, the costs that pesticides pose to the environment and human health were not so easily quantified. Regulatory practice to some extent consistently underestimated the costs of each pesticide because it simply lacked the information needed to make such determinations. When the EPA's Office of Pesticide Programs was created, it essentially had no reliable data with which to evaluate the costs of pesticides already in use, as little to no toxicological or other scientific research had ever been conducted to evaluate their health effects.[51] This lack of supporting data played a significant role in the office's astoundingly slow progress in evaluating the human health impacts of previously registered pesticides. At the same time, the EPA was able to keep registering new pesticides relatively quickly through a "conditional registration" process that speeds the review of purportedly less toxic and other high-priority pesticides.

The structure of risk assessment itself—the primary tool that regulatory agencies use to assess and regulate pesticides—also helps to explain the U.S. EPA's ease with registering new products and its difficulty restricting pesticides already in use. As previously discussed, risk assessment is the process through which scientists examine individual pesticides acting in isolation and then quantify the probability of harm to human health associated with varying levels of exposure. But the process routinely underestimates the health risks posed by pesticides because of its reductionist analysis of isolated chemicals, widespread gaps in scientific data, and risk assessors' reliance on scientific norms that dismiss considerable data. Carolyn Raffensperger and Joel Tickner are among a growing group of scholars who critically evaluate the structure of risk assessment. They point out that risk assessment is a poor fit for many fields in which it has become institutionalized:

Risk assessment, which was originally developed for mechanical problems such as bridge construction where the technical process and parameters are well defined and can be analyzed, took on the role of predictor of extremely uncertain and highly variable events. The risk-based approach, now central to environmental and public health decision making in the United States, has in part led to a regulatory structure based on pollution control and remediation, rather than fundamental prevention.[52]

Although the EPA can require pesticide manufacturers to submit evidence of safety for new products, the structure of risk assessment itself tends to minimize the risks associated with pesticides, contribute to the dominance of the pesticide paradigm within agricultural pest management, and bolster regulatory agencies' propensities to regard pesticides as posing minimal risks (and offering relatively high benefits) and accordingly register them for widespread use with minimal regulatory controls. Scientists within and beyond the regulatory arena have responded to some of these critiques in recent years, such as by incorporating greater safety margins where the data are lacking or inconclusive, designing toxicological studies to identify more sensitive health end points (e.g., watching for eye irritation rather than death), and conducting research on hitherto rarely studied health impacts.

Yet it is important to recognize the difference between what risk assessors are learning to do to improve their work and what gets done
in practice within the regulatory arena. Regulatory agencies frequently
make deliberate decisions throughout the processes of creating as well as
using risk assessments that inexplicably deviate from otherwise-standard
scientific protocol and are unjustifiably industry protective. Outside observers and regulatory agency scientists alike often criticize the EPA and
DPR risk assessments for relying on shoddy toxicological studies, ignoring
published peer-reviewed studies, defaulting to a less health-protective option in cases of scientific disagreement, arbitrarily excluding some data,
ignoring entire exposure pathways (e.g., inhalation) along with entire
classes of human health impacts, assuming that a lack of data indicates
zero risk, underestimating the degree to which genetic variation and a
disproportionate risk of exposure can make some human populations
much more susceptible to illness from exposure than the "average" adult,
using models that assume that no prevailing wind exists and thus that the
pesticide disperses equally in all directions, failing to include widely accepted health-protective uncertainty factors, and obscuring many of the
above problems.[53] Of particular relevance to the issue of pesticide drift,
the U.S. EPA only assesses the potential of exposure via inhalation for
highly volatile pesticides and assumes that exposure is zero for all "semi-
volatile" pesticides. California regulators and scientists at PAN, however,
have found dangerous concentrations of semivolatile pesticides in the air
outside agricultural fields, and as such, understand that such pesticides
present inhalation exposure hazards to "bystanders." The EPA also relies
extensively on health studies supplied by industry. Congress and the EPA's
own scientists have criticized the agency consistently since at least the
1960s for not verifying the validity of those industry data, and industry
has been found to have withheld data showing that particular chemicals
under review (and in use) pose serious health hazards.[54]

In some cases, regulatory agency management simply ignores its own
scientists' risk assessments and instead bases risk management decisions
on arbitrary goals that require fewer regulatory restrictions. DPR's regulatory decisions about methyl iodide are a case in point. In addition to
the fact that the agencies' risk assessors made numerous errors that jeopardize human health (including relying on shoddy studies commissioned
by the manufacturer, ignoring crucial health effects, and making unrealistic assumptions about the efficacy of worker protective equipment),
DPR management proposed to regulate methyl iodide to a health benchmark that was inexplicably 120 times higher (i.e., less health protective)

than the level recommended by its own risk assessment staff scientists. In their scathing criticism of DPR's actions at a legislative hearing in June 2010 on those proposed regulations, two scientists serving on the DPR-commissioned methyl iodide Scientific Review Committee stated,

In the end, the scientists at DPR made estimates of target levels [health benchmarks] of methyl iodide available to the public and to the science review committee for comment in 2009 and incorporated many if not all of these comments in the final February 2010 human health risk document. Many scientists feel that these values still underestimated the health risks posed by MeI. However, these peer-reviewed, "speed limit values"—not to be exceeded—were ignored by the risk managers, who developed their own numbers without review by either the public or the science review committee. These target maximum values are over 100 times higher than those developed by DPR's own scientists and that have undergone extensive peer review.

It seems specious to me to even to begin to try to calculate something that will mitigate and make something safe when you have no idea where the No Adverse Effect Level [health benchmark] is, except the certainty that it is much lower than anything that we have already looked at.

Risk assessments that routinely underestimate the risks associated with pesticides lead to pesticide regulations that do not adequately protect public health, thereby translating to problematic pesticide use patterns and widespread public exposure to agricultural pesticides. The regulatory processes of conducting and using risk assessment were made subject to extraordinary criticism for precisely these political controversies and technical limitations throughout the 1980s. Such critiques were leveled at nearly every aspect of the EPA's work by industry, environmentalists, and Congress alike. Sheila Jasanoff, who has long observed regulatory science, suggests that this critique effectively ended regulators' hopes "to shelter their discretionary judgments at the borderline of science and policy behind uncontested claims of scientific expertise."[55] As a result, the U.S. EPA incorporated substantial transparency and peer review into its decision-making processes. These efforts to bring transparency and accountability to the risk assessment process and to separate scientific analysis from political judgment typically led to new forms of paralysis, where significant regulatory action only resulted for the small number of organochlorine pesticides that were widely recognized as environmentally problematic (like DDT).

Notwithstanding the critiques of risk assessment and its use, it is now widely institutionalized as the dominant approach to evaluating and regulating pesticides. It forms the core of pesticide regulatory practice, which pivots around licensing, controlling, capturing, and otherwise *managing* pesticide pollution. Moreover, as Joe Thornton argues, it is generally perceived as the only reasonable, logical, and objective approach, and its adherents do not recognize that other approaches are possible. Thornton therefore conceptualizes risk assessment as not simply a tool but rather the central component of the "risk paradigm"—"a total way of seeing the world, a lens that determines how we collect and interpret data, draw conclusions from them, and determine what kind of response, if any, is appropriate."[56] The following statement from a DPR scientist, in response to my interview question about environmental advocates' critiques of DPR, illustrates this ideological, worldview quality of risk assessment:

I think sometimes it just comes down to the fact that we don't see the world the same way that these folks are seeing it and they don't understand why we don't see it that way. Therefore [they think that] we must be influenced [by industry], because if I had the information they had, I must feel the same way they feel. And I don't. Because I have all of this other information that says, yeah, there's a risk there, but there's a risk to do with this and with that.

Because risk assessments systematically underrepresent the threats posed by toxic chemicals, the worldview of the risk paradigm effectively legitimizes the pesticide paradigm. More concretely, the risk paradigm's approach of chemical-by-chemical assessment and control suits the needs of industry well, as it justifies regulatory restrictions only for those specific pesticides for which substantial scientific research has unequivocally demonstrated harm.

The building blocks of the major legislative and regulatory changes of the 1970s, in sum, thoroughly embodied the pesticide paradigm, which was further legitimized within the regulatory arena by the predominance of the risk paradigm and its limited tools, solidified by regulatory agencies' restrictive legislative mandates, and lead to and legitimized minimal regulations of pesticide use. As a result, regulatory agencies have fulfilled one long-standing function of pesticide regulation—to bring new products to market—and have been generally unable to restrict pesticides already in use.

Funding

The U.S. EPA's failure to evaluate the safety of pesticides in use can also be partially attributed to inadequate funding. Budget authorizations to the U.S. EPA plummeted sharply after the agency's inception and have not increased since that time (see figure 4.1).

Such funding declines are striking, given the enormity of the task that the agency had been assigned and the fact that its responsibilities have

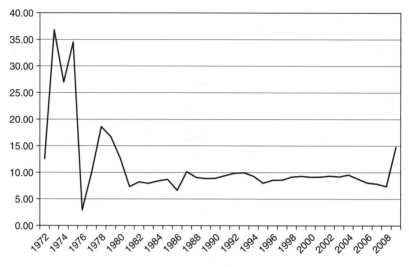

Figure 4.1
Budget authority for the U.S. EPA, 1972–2009. *Notes*: The EPA was established in 1972. The values have been adjusted for inflation and presented in 2009 dollars. *Source*: U.S. Office of Management and Budget 2010.

increased over time. As with other work that the EPA conducted, the agency's pesticide program in the 1970s and early 1980s was widely critiqued as expensive as well as ineffective. Congress controls funding allocations for federal regulatory programs and responded to these critiques of regulatory ineffectiveness by cutting back on the agency's funding. Working to fulfill an overwhelming legislative mandate in a context of inadequate and declining funds, the U.S. EPA's Office of Pesticide Programs sped up the assembly line and cut corners to keep up—registering new products as quickly as possible and conducting shoddy evaluations of those already in use.

Budget reallocations continue to compromise the U.S. EPA's ability to fulfill its mandates. In 2006, the Bush administration cut the agency's overall budget, targeting programs that prioritize children's health protection. At the same time, the administration moved forty-five million dollars out of EPA research funds to create new national security functions that link the EPA to the Department of Homeland Security.[57] Budget cuts in 2008 reduced funding for EPA libraries and databases that provide publicly available scientific information about toxic chemicals. This information is widely used by local, state, national, and international agencies, and ten thousand EPA scientists and other staff vociferously opposed the closure of the libraries.[58]

In California, the reduction of pesticide regulatory funds occurred much more recently, when DPR suffered serious budget cuts in 2001–2002 that caused staffing cutbacks.[59] Particularly hard-hit in the budget cuts of that year and of considerable relevance to the issue of pesticide drift were the state's risk assessment capacity (cut by one-third) along with its already-starved air sampling and analysis funds (cut by 60 percent).[60] These minimizations of state pesticide regulatory capacity also limit DPR's ability to oversee CACs' enforcement of pesticide use guidelines and response to reports of pesticide drift and exposure. DPR's own discussion of the cuts makes it clear that they exacerbate the problem of pesticide drift:

The reduction scales back efforts to develop analytical methods for fumigants and other pesticides and delays the initiation of mitigation for pesticides listed as toxic air contaminants. Environmental fate descriptions for risk assessments and conducting field studies to quantify public exposure to pesticides will be delayed.[61]

In 2009 and 2010, California faced a new budget crisis that posed grim implications for pesticide drift. During times of budget cuts, resources get stretched thin; all California EPA staff members were required to accept a 15 percent furlough (cut in work hours and salary). Moreover, it can be assumed that the fear of job loss compelled some individual staff members to become even less willing to critique economically valuable pesticides.

Industry Influence
Regulatory inaction and the pervasiveness of pesticide drift must also be understood as stemming from the tremendous pressure that the agricultural and chemical industries continue to exert on environmental regulatory agencies. Investigative journalists and other researchers have shown how industry uses various direct and indirect tactics to shape policy, regulations, regulatory practice, and legal processes in its favor, and avoid liability for illness and injury.[62]

In the legislative arena, the agricultural and chemical industries have always dominated decision making at the state and federal levels, helping to reduce budget allocations to environmental agencies and limit environmental reform in both California and the U.S. EPA. Congress made innumerable proposals to strengthen pesticide regulation over the years, yet Bosso shows that "the ability of farm bloc [legislative] representatives to use their positions as bulwarks against policy modification" meant that legislators pursued change "without surrendering their core prerogatives" of economic growth and industry protection.[63] Where legislation has been passed that increases the U.S. EPA's pesticide regulatory responsibilities, it is typically geared toward protecting consumers from exposure to pesticide residues on food—not protecting agricultural community residents from pesticides in air and water.[64] The farming and chemical industries spend extraordinary resources every year directly lobbying congressional representatives, greatly outpacing the relatively meager resources of donation-dependent environmental advocacy organizations. The legislative protection of agricultural industry interests is bolstered by cultural beliefs that are widespread in the United States—notably, the association of U.S. agriculture with an idealized image of the family farm, and the idea that family farms represent the bastion of U.S. virtue, democracy, and community values. These notions dominate legislative debates about agriculture and undermine legislative proposals to restrict the technologies on which so much agriculture currently depends.

Regulatory agency work is similarly shaped by industry interests. As discussed in chapter 3, a revolving door continues to exist between industry and regulatory management, ensuring that regulatory decisions will never stray too far from industry interests. Pesticide regulatory leadership in California is politically appointed, and DPR itself boasts that its top officials always hail from industry. For example, the current DPR director, Mary Ann Warmerdam, spent twenty years working for the California Farm Bureau Federation, the major growers' lobbying organization in the state. DPR's current chief deputy director was the executive director of the Manufacturer's Council of the Central Valley—an industry organization—before joining DPR.[65]

Industry influence is also built into the statutory framework within which regulatory agencies were constructed. For instance, when the California EPA was designed in the early 1990s, authority over the assessment and regulation of pesticides was allocated to DPR, and DPR's staff members were pulled primarily from the California Department of Food and Agriculture (an institution charged with promoting and protecting agriculture within the state). The fact that DPR was given control over evaluating pesticide risk is noteworthy, since the Office of Environmental Health Hazards Assessment (OEHHA), which was comprised of "public health-oriented toxicologists" from the Department of Health Services, conducts risk assessments for all other chemicals. This institutional design was done to explicitly appease the agricultural industry, which felt threatened by the prospect of its practices being regulated by an environmental agency. The relationship between these two California EPA institutions (DPR and OEHHA) is not clearly defined. Although DPR asserts that it is only mandated to consider OEHHA's input in risk assessments, OEHHA and other institutions argue that DPR's regular dismissal of OEHHA arguments is a violation of regulatory statute. The dispute is not inconsequential, since DPR's risk assessments and associated regulatory decisions are routinely less health protective than the recommendations that OEHHA has made.[66] One California EPA agency insider observed in an interview with me,

It really becomes very disconcerting when the number two person in Cal/EPA has always been an agriculture person, and that has always resulted in a bias toward protecting the interests of the DPR when it comes to pesticides and ignoring the OEHHA role. . . . As it currently stands in 2006, OEHHA has an extremely small part in regulating pesticides in the state of California. . . . If you were to actually pick five chemicals out of a hat and have DPR do an assessment and OEHHA do an assessment, you would get different results, with OEHHA's being more protective.

Although some regulatory agency staff members insist that industry has no more influence over regulatory practice than do environmental or public health advocates, numerous investigations indicate otherwise. Agency scientists interviewed under conditions of confidentiality report that industry influence over their work is disproportionate, significant, pervasive, and oppressive. Procedural changes made during the Bush administration shifted decision-making power away from U.S. EPA scientists and increasingly into the hands of politically appointed officials.[67] During that time, many regulatory scientists publicly announced that their work was actively constrained by the political interests of top agency officials.[68] In 2006, for example, union leaders representing thousands of U.S. EPA scientists publicly declared that industry influence within the agency directly leads to shoddy risk assessments and inadequate pesticide regulation:

Our colleagues in the Pesticide Program feel besieged by political pressure exerted by Agency officials perceived to be too closely aligned with the pesticide industry and former EPA officials now representing the pesticide and agricultural community. . . . [The prevailing] belief among managers in the Pesticide and Toxics Programs [is] that regulatory decisions should only be made after reaching full consensus with the regulated pesticide and chemicals industry.[69]

Other U.S. EPA scientists interviewed on the condition of confidentiality by the *New York Times* during that period made similar accusations, indicating that the major legislative and regulatory reforms of the 1970s have done little to change the character and consequences of pesticide regulation:

It's how the game is played. . . . You go to a meeting, and word comes down that this is an important chemical, this is one we've got to save. . . . The pesticide program functions as a governmental cover for what is effectively a private industry licensing program.[70]

Under the condition of confidentiality, one former senior California EPA official spoke frankly with me about the ways in which politics and industry interests similarly shape pesticide regulatory science in California:

Folks at DPR who do the risk assessments, for example, aren't the ones who make the management decisions usually. Usually it's the upper management, and sometimes it trickles back down so that they [the scientists] have to change their numbers in order to meet some goal of the department. . . . It's politics. I know this happens.

They have the science arm thing, but then the information gets floated up to the upper management levels, and they start considering, as their mandate is, they start considering the benefits, the feasibility, whatever, and they start changing the risk assessment so there isn't this big difference between what the risk assessors are saying the numbers should be and what the managers [say it should be]. So the final number is just one unified number that comes out of DPR, and it looks like it's based on science. . . . This happens at U.S. EPA too. This is why it's broken over there too. Because they say, we need it to be this number so we can continue to use this. You fix your risk assessment so it meets that number. We don't want to have it be this big open clearly articulated difference between the science and the final [number]. We want it to look like they are closer together. So they have to fudge numbers.

Numerous DPR scientists who I interviewed stated that they work under little industry pressure, and a few even pointed out that some industry pressure is beneficial because it forces them to carefully justify their decisions. That said, these scientists also recognized at other points in the interviews that compared to environmental or public health advocates, industry is disproportionately able to scrutinize and weigh in on DPR's work. Though scientists may have relative freedom from industry influence within the risk assessment process, upper management (which does interact directly and frequently with industry) has tremendous discretion over how and to what extent risk assessments will ultimately shape pesticide regulations. Scientists' risk assessments often flounder for years in the hands of managers before risk mitigation strategies are debated, devised, and then finally implemented as guidelines or regulations. These issues emerged in an interview that I conducted with a regulatory scientist who had, at that time, just recently completed an elaborate risk assessment on a pesticide that is highly toxic to humans and whose (already-high) use is anticipated to increase as its primary competitor is phased out due to environmental concerns. Toward the end of our long interview, this scientist expressed concern about the degree to which upper management would take seriously the risk assessment he completed:

The real question [is], are they [management] going to act on [the risk assessment]? Are they going to base mitigation on it? . . . I want to put out as good a risk assessment as I can. That's what I'm really here to do. And I've written things here and advised them on things, and whether they listen or not, I don't know. But I'm, I have high concerns about this chemical, knowing full well that it could be a replacement for [its main alternate pesticide], knowing full well that its use went up and up through the '90s. . . . When [the alternate pesticide] goes, this stuff is I think right in line. So . . . its use might very well go up, and if we still have situations where [ambient concentrations of the pesticide exceed the health benchmark, as they observed], we're going to have lots of problems out there. It's not fair to those people out there, frankly. [pauses] I should not have said that, because I'm supposed to be a scientist.[71]

Industry influence over pesticide regulatory work does not always take direct, overt forms. Because industry resources greatly outnumber those of environmental and public health advocates, industry groups have a much greater ability to actively contribute to specific scientific debates, mitigation deliberations, enforcement actions, or legislative proposals. Also, regulatory agencies intentionally consult with industry throughout the regulatory process (and typically before other interest groups like public health and environmental advocacy organizations) in order to avoid costly legal battles that might develop if industry actors perceived that they were excluded or that the agencies were otherwise being too contentious. In other words, the threat of lawsuits from industry effectively forces regulatory agencies to negotiate with industry regularly, and industry's extraordinary resources make that threat a legitimate one. Industry representatives also often serve advisory roles in notable "neutral" scientific institutions, including the National Academy of Sciences and the National Research Council.

Chemical industry organizations also secure minimal regulatory restrictions by obscuring the dangerous impacts of their products. They do so by hiding pesticide components from public view under the guise of protecting "proprietary knowledge," hiring public relations firms, contracting scientists to conduct and disseminate scientific studies that create uncertainty over what would otherwise be clear health risks, refusing to publish the findings of studies they have funded where those results show pesticides to be unsafe, and investing in and otherwise manipulating the media.[72]

Regulatory Culture

As much as trends in regulatory practice stem from the material factors discussed thus far (i.e., legislative mandates, funding levels, and industry pressure on agency officials), regulatory work is also a cultural phenomenon—shaped by beliefs, ideas, and norms. Regulatory culture shapes the ways in which regulators conceptualize and thus try to solve issues like pesticide drift. I have already noted several cultural phenomena that contribute to the pesticide drift problem, including the widespread social notions of progress that legitimize the industrial model of agriculture, scientific norms that constrain the health-protective capacity of risk assessments, and the widespread belief in risk assessment as the only appropriate way to evaluate the safety of toxic chemicals.

Two pervasive and influential aspects of contemporary pesticide regulatory culture that deserve a moment of focused attention are those of industry protection and the pesticide paradigm. Researchers have clearly established that both of these were historically important factors shaping early pesticide regulation, the initial purpose of which was to protect farmers and their crops from defective pesticides as well as bring new, effective pesticides into the marketplace. Because industry protection and the pesticide paradigm were deeply ingrained in the culture of the USDA, the USDA staff members who were transferred into the U.S. EPA's Office of Pesticide Programs in the 1970s carried this regulatory culture with them into their new institution. Through interviews with early Office of Pesticide Programs staff members, Bosso shows that this office's managers were sympathetic to industry needs and committed to minimal regulatory controls on pesticides.[73] Bosso contends that this played a major role in the agency's early failure to implement FIFRA, secret deals with industry, and systematic exclusion of the public from the agency's decision-making processes.

Certainly, U.S. and California EPA staff members today differ from their earlier counterparts, as concern for the environment that gave rise to the environmental movement similarly shapes the convictions and commitments of many who work in regulatory agencies. All the individuals I interviewed expressed considerable care about the impacts of agricultural pesticides on the environment and human health. Notwithstanding these environmental concerns, the culture of protecting chemical-intensive agriculture remains alive and well within many departments of pesticide regulatory agencies, where alternative models of agricultural pest management simply are not considered to be realistic or legitimate alternatives to the pesticide-intensive model.

The culture of industry protection and the pesticide paradigm are particularly evident at the county level. Regulatory culture at the local level plays a large role in shaping the extent to which pesticides are regulated and the forms that regulatory presence takes, since California's CACs are the primary enforcement agents for state pesticide laws and regulations. For example, CACs have shown a clear reluctance to discipline growers and pesticide applicators in regulatory violation cases. DPR data indicate that from 1998 to 2007, CACs statewide were eight times more likely to issue a warning letter than a monetary fine in instances of regulatory violations, and in Tulare County the average annual number of warning letters was twelve times higher than monetary fines. These enforcement track records are also significant in absolute terms: in even the state's agricultural powerhouse counties (Kern, Tulare, and Fresno)—where hundreds of thousands of pesticide applications take place every year—CACs annually issued only an average of twenty-two fines for pesticide regulatory violations.[74] Moreover, the average fine for a pesticide violation is only a few hundred dollars.[75] Although regulators recognize that the threat of fines compels growers to more carefully comply with regulations, they still emphasize the value of flexibility in dealing with violators and express a reluctance to regulate. In an interview with me, one CAC discussed his preference for flexibility with issuing fines and revealed a strong sympathy for industry in doing so: "[People have] heard me say this a million times: a sound regulatory scheme differentiates between willful and negligent violators and people that are making a good faith effort to comply with this *huge* body of regulations that they have to comply with."

Local regulators also demonstrate a clear reluctance to restrict pesticide use, and a belief in the necessity and safety of the chemical-intensive, industrial model of agriculture. Though California regulators widely stress that considerable pesticide regulatory power rests at the local level, CACs in general underutilize their authority to regulate pesticides any more than the bare minimum. CACs, for example, have the authority to implement permit conditions for nonrestricted materials, yet this is only rarely done (in spite of the fact that many nonrestricted materials are implicated in the pesticide illnesses that show up in DPR's PISP database). Additionally, DPR and CACs rely on informal, flexible pesticide use guidelines (rather than regulations), so that each specific application scenario can be tailored to the local conditions. This preference for flexibility has a logical basis, as it helps CACs tailor their activities to the considerable variation in landscape, cropping patterns, climate, and population pressures that typically characterize California counties. Still, where CACs are highly protective of industry, this flexibility becomes a loophole that effectively serves the interests of growers. Even DPR's deputy director admitted that many permit applications for restricted material pesticides are less protective than the recommended statewide guidelines.[76] Although the public can technically appeal any permit application, the process requires extraordinary vigilance and time, is therefore rarely utilized, and in turn effectively protects the growers and pesticide applicators who submit them.

Local regulators defend this devolved, flexible approach to pesticide regulation. After elaborating on the value of informal guidelines rather than formalized regulations, one deputy CAC asserted that county officials should maintain authority over pesticide regulations because formalized EPA regulations "punish the whole industry." At the end of our interview, this person noted that she wanted to make two last points: the public needs to understand that pesticides are necessary in life, and pesticide drift is not "the problem" as much as the claim that "all pesticides are bad" is.

The adherence to the pesticide paradigm is also evident in the narrow ways that CACs conceptualize alternatives to problematic pesticides. For example, in response to residents' concerns at a public hearing about proposed regulations for methyl bromide in terms of the pesticide's human health impacts, a deputy CAC tried to placate the crowd by noting that "alternatives" were being developed, "like Inline." What he did not mention is that Inline presents equally significant risks to the health of people living near agricultural fields, since it is a combination of 1,3-dichloropropene (a known carcinogen) and chloropicrin (carcinogenic and highly acutely toxic). CACs do not give much attention to alternative, ecological models of agricultural pest management, as evident in the following statement made by one CAC in an interview with me:

I've told people in industry. . . . You have to really get out there about the importance of food. That this country really needs to be able to provide for itself, whether it's energy or food. . . . There is a need for us, whether it's within our homes, to treat for pests or for us to have enough food to feed ourselves and the world. There's no way [of] getting around it. And the thought of, well, let's get away from pesticide: well, what does that force researchers to do? Let's come up with a plant that can fend off—well, now you're into a whole other area that people don't like, and that's manipulating genes.

In this framing, current pesticide use patterns are presented as essential to food production, and GM crops constitute the only alternative. Such claims function ideologically, as they represent current pesticide use patterns as natural, necessary, and legitimate.

CACs emphasized the importance of a collaborative approach to working with growers. One deputy CAC elaborated at length about the value of "building relationships with industry" as the best way to enforce pesticide regulations. This CAC representative underscored that these "collaborative" approaches with industry can be effective, and that otherwise industry would tie regulatory agencies up in court. Public health advocates and other observers of local pesticide regulatory enforcement look at this in a different light, suggesting that the "cooperative" approach to enforcement disproportionately protects local growers in ways that downplay and dismiss the consequences of pesticide drift. As one farmworker advocate observed in an interview with me, "When people call in to report being sprayed, ag commissioners often tell the caller that they will go talk to the sprayer instead of filing an official report."

In interviews with me, many CACs' descriptions of their various responsibilities indicated that they perceive growers as their primary constituents, and that helping growers is their principal and most gratifying responsibility. For example, even though I made it clear that my research interests were about the human health impacts of pesticide drift, most of the county regulators I interviewed elaborated at length about their efforts to protect crops from drifting herbicides as their most successful work to control the pesticide drift problem. This is understandable given that drift on to neighboring crops is a less controversial issue, since the solutions benefit all parties (all growers have crops with the potential to be drifted on). One CAC representative I interviewed lamented that such accomplishments are not adequately recognized: "The success of what we've achieved on the west side [of the valley] with [reducing crop-damaging] drift is *huge,* and that story isn't being told."

In describing to me their work and the different groups with which they interact, CAC representatives consistently characterized growers in respectful terms—as reasonable, rational, sympathetic, selfless, and well intentioned:

I think that growers who are in these situations try to be good neighbors.

The growers give schools a wide berth, too. It's incredible what they do sometimes.

Our applicators, especially the aerial applicators, will call us and say, "I'm not feeling real comfortable about this. Can your staff come out and take a look? And if you deny the job, we're perfectly fine with that."

What I have found with the ag industry, you just tell them what it is they need to do. "Just tell us what we need to do. Let us know you were fair in the assessment, and that's what we'll do." That's been my experience the last ten years or so. . . . Now pretty much the attitude is, "Yeah, we'll find a way"—not to get around the regulation but a way to comply or adjust traditional means of farming. But there will come a point where even that's not good enough anymore, and I would understand somebody just throwing up their hands and going, "Forget it! I can't make you happy!"

By and large I think they do a good job. We hear about the ones that we take enforcement action on, but there are a great many of them that take place that are very, very, very, very conscientious and very safe, and you never hear about them because it doesn't require us to do any enforcement. . . . The overwhelming majority are very conscientious applicators. . . . One overwhelming thing, from our perception, is that *most* growers are very aware of the materials that they are using, very aware of their surroundings, and use those materials responsibly and safely. . . . You don't hear about the people that are trying to do the right thing and a good job.

In contrast, CACs often expressed frustration about public health advocates and people who suspect that they have been exposed to pesticide drift. Although they point to the good relationships that they have developed with many environmental advocates, CACs also express disappointment about having to spend time justifying their work to people who do not necessarily recognize CACs' and growers' accomplishments. CACs universally refer to residents' concerns about pesticide drift as "complaints" (rather than "concerns") and emphasize that many "complaints" are simply unreasonable, stemming from what they see as irrational, uneducated, or uninformed anxiety. The following statements made by local regulatory officials in interviews with me exemplify the condescending and exasperated tones that regulators frequently take when describing residents' concerns about pesticide drift:

I think there are some circumstances that we run into with more urbanized environments where people who don't have knowledge of agriculture, or familiarity with the processes, are maybe somewhat confused or offended or taken aback by them. I can many times help people to better understand the risks that they have or the significance of what their concern is based on a little basic biology. I mean mold is a perfect current example of that. You know the sensitivity in the media of mold and yet, you say, well, you known there is some good molds . . . and yet we got this anxiety created.

When I get these calls, they've already got it in their minds: pesticides are bad. So anything that you may say to them to try to counter what they've read on the Internet from whatever source . . . they're not being open-minded.

Where does it stop? . . . Anything can be toxic. Anything. So where are you going to draw the line and say this is acceptable? . . . I can't be around somebody who's using certain perfumes! Why is that OK? That's a chemical! But pesticides seem to be the easy one to target. . . . There is also chemiphobia out there. . . . Pesticides seems to be the boogieman.

[Some activists] said, "We'd like you to absolutely ban all pesticide use within a perimeter of a house, all houses." In that meeting, we said, "Well, what would you do if you had cockroaches or ants in your garage, what would you do?" "Well, we'd spray them." . . . Some of the people who ask for this prohibition were the ones telling me that if they had ants they would spray them. . . . I know my wife gets really excited when we find ants in the house. She hates that.

At regulatory hearings and other public events, I too have witnessed some of the phenomena that the CACs express frustration with: the activists who insist that "all pesticides are bad," refuse to accept any CACs' assurances of safety, attribute seemingly every personal problem to pesticides, and frame their work as an effort to prove what they already know to be true (rather than to test their suspicions). Many reports of suspected pesticide drift incidents are in the end false alarms—strong odors and visible drifts from airplanes sometimes turn out to be only seeds, fertilizer, or water.

That said, these behaviors do not fairly characterize the work of most pesticide drift activists, who generally express a more nuanced understanding of variations in toxicity among pesticides and spend time politically addressing many other environmental contaminants (not just agricultural pesticides). Also, the apparent errors that activists make about suspected pesticide drift incidents should be understood as stemming from the unavailability of clear evidence of safety and the indisputable fact that a foul odor might indicate exposure to a highly toxic chemical. Activists are often unwilling to accept CACs' placations because most of the time, there is simply no way to ascertain the accuracy of their suspicions. Similarly, the mistakes that activists do admittedly sometimes make about various chemicals' toxicities generally come from the activists' lack of formal scientific training rather than simply from a determined refusal to consider new information. CACs' nearly universally disparaging depictions of activists thus are striking and notable, particularly in contrast to their expressed respect and understanding for growers and other pesticide users.

The regulatory culture of protecting growers and chemicals emerges from multiple material factors. Unlike the California EPA's DPR and the U.S. EPA's Office of Pesticide Programs, which were moved out of agriculture departments and designed, at least ostensibly, to be environmental protection institutions, CACs are tasked with two objectives that are frequently at odds: promoting agriculture and protecting the environment.[77] Indeed, during interviews with me, multiple agriculture commissioner representatives rattled their mission statements off to me unprovoked, emphasizing repeatedly their "duty" to promote the agricultural industry. Through interacting regularly with growers, moreover, agriculture commissioners and other regulatory officials understand that restrictions on pesticide use often increase labor costs or pose other economic hardships in the short term, especially for small or otherwise relatively marginalized growers. Additionally, because the agriculture commissioners are politically appointed by local boards of supervisors, local pesticide regulatory practice tends to reflect the interests of local elites: economic growth in general and extra protection of agricultural interests in counties with particularly strong agricultural economies. Because CACs have tremendous regulatory authority and discretion, the nature of regulatory culture at the county level in California plays a major role in making pesticide drift the problem that it is today.

Telling Stories: Obscuring the Problem and Naturalizing Regulatory Response

While the material and cultural factors discussed above have helped produce the pesticide drift problem, the primary stories told about the issue help to make it seem unproblematic. Stories or narratives are a primary mechanism through which people frame a problem, giving it a specific texture and character. In any given story, some aspects of the issue are brought to the foreground while others are left out. Narratives about a problem give it shape, identify its size and scope, define its causes, and point to the appropriate solutions. The dominant narratives told about pesticide drift do some powerful work, as they cast regulatory response in a particular light: as reasonable, natural, and unproblematic. Specifically, regulatory officials frame pesticide drift as a series of localized, isolated accidents in an otherwise-effective regulatory system.[78] In this section, I show how this narrative framing is constructed in California, where regulatory agencies perform a lead part in public conversations about pesticide drift.

First, DPR representatives point to regulatory data on pesticide illnesses to frame pesticide drift as a rare and small issue. In news reports, press releases, and interviews about pesticide drift, regulatory officials consistently draw on DPR's official pesticide illness data and stress their insignificance relative to the frequency of pesticide applications. As the then-director of DPR stated in a newspaper article, "There are over a million pesticide applications every year [in California]. The incidents of drift are around 40 per year, so that's a relatively small number."[79] DPR officials faithfully adhere to this narrative, as evidenced by the following statement made by DPR's director in late 2009: "If you take into account the thousands of applications that occur in California every year, we still have a remarkably compliant agricultural sector."[80] Every CAC I interviewed similarly highlighted the value of contextualizing illness data, as in the following statement:

I think it is important to put in context when adverse incidents happen. If you have, let's say you have twenty adverse incidents in a county in a year. That could sound like a lot. You could say, "Wow. County XYZ had twenty illnesses." That needs to be viewed in the context of [the fact that] some counties would have fifty to sixty thousand applications. So, is it too many? Of course it's too many. Nobody should ever be made ill by an ag chemical application. But you almost never hear that put in context. Almost never.

To justify these claims that the problem happens only rarely, DPR points to the reliability of its data collection systems. To be fair, DPR occasionally recognizes the limitations of its PISP system.[81] The prevailing narrative from DPR and CACs, however, is that these data fairly characterize the scope of the problem. For example, the press release for its PISP data for 2002 stated, "DPR's program is very effective at detecting any incident involving multiple victims."[82] The summary statement for 2003 likewise asserted that "DPR maintains a high degree of confidence that the Pesticide Illness Surveillance System captures the majority of agricultural pesticide illnesses."[83] In interviews with me, CAC representatives noted the validity of official data by pointing to the numerous mechanisms through which the public can report pesticide drift and illness. For example:

I look at the thousands of applications that occur in this state. *Thousands!* Structural, maintenance gardeners, homeowners! And gosh, I don't know how many more venues there can be to have people report to us, especially in this day and age with cell phones and the Internet. . . . Where's the data to say that people get exposed? There are thousands of pesticide applications every year. There are more than enough venues for reporting pesticide exposure.

Each year, DPR releases the most recent batch of data on pesticide illnesses in the state, and its narrative accounts of those trends similarly minimize the problem. DPR officials' statements to the press invariably attribute declines in the numbers to actual declines in pesticide illnesses, whereas they attribute increases in the illness numbers to improvements in data collection procedures.

Regulatory officials draw on narratives of scientific rationality and objectivity to further validate official data of pesticide drift and illness. Regulatory scientists and managers insist that they will accept "any data that pass scientific muster," emphasize that the protocols they follow reveal "the truth," and typically dismiss input from environmental activists as "opinions."[84] In an interview with me, one deputy CAC alleged that DPR deviated from these scientific norms when it created stronger enforcement response regulations: "Regulators [at DPR] are responding because of the political pressure. And I don't mind that. But then where's the science? Where's some honest science to back that up and justify it?" This individual is underscoring the notion that science can work as a neutral arbiter that weeds out weak or politically biased information.[85] By defining an issue as belonging in the realm of science rather than politics, this speaker is attempting to remove the issue from public debate. Such narratives also obscure the data gaps, industry privilege, corruption, and other material factors that consistently compromise science in ways that effectively minimize official assessments of pesticide drift. The discursive invocation of science thus further legitimizes regulatory framings of the problem as minor and justifies the minimal regulatory response to pesticide drift. As I will show later in this chapter, activists problematize these claims by emphasizing that scientific facts are contingent on and shaped by technical limits, social inequalities, and normative decisions.

Furthermore, officials characterize incidents as "accidents": the result of pesticide applicator error. In interviews with me and in public statements, CACs, county supervisors, and DPR officials consistently stressed this accident narrative:

It's like airplane crashes—it's operator error.

We do have the human condition to contend with, and mistakes do happen.

It is the people who are not following the rules who are creating the problem.

The material fumes and will move under certain conditions if it is not applied correctly.

The system works. . . . Unfortunately, we have people who don't follow the law.[86]

Officials contend that *careless* people who deviate from the norm make these errors. For example, a DPR spokesperson justified the department's particular response to pesticide drift by noting, "We don't want the reputation of the industry ruined by a few bad apples or careless acts."[87] Framing pesticide drift incidents as being relatively few in number as well as limited to applicator error enables officials to argue that the system is effective and that no new rules are needed. As one CAC stated in an interview with me, "When you have a relatively small number of incidents that impact the people or the environment, the system for the most part works."

Clearly, violations of specific pesticide regulations warrant increased enforcement of existing regulations and continued applicator training. To a large extent, though, the accident narrative is simply inaccurate. As discussed earlier in this chapter, DPR's own case files show that nearly half of all illnesses due to pesticide drift are not attributed to *any* regulatory violation. These findings suggest the need for increased controls on pesticide use and more precise regulations. Moreover, the accident narrative is problematic for the ways that it obscures the inherent uncontrollability and unruliness of drift-prone pesticides.[88] Yet the accident narrative dominates the discursive landscape, and it does so for many reasons. Elected and appointed officials repeatedly assert the accident storyline and ignore evidence of its inaccuracy, thus reinforcing its apparent validity. The numerous social inequalities that prevent many pesticide illnesses from making their way into official reporting systems also reinforce the claim that pesticide drift events occur only rarely. Then too, the most common violations cited in drift cases are simply the allowance of drift to happen and the failure to use due care; consequently, many pesticide illness events otherwise unattributed to regulatory violations are characterized as errors or accidents. In light of local pesticide regulatory officials' demonstrated sympathy for growers, it does not seem unreasonable to suspect that official investigations of pesticide drift events may be biased in favor of growers and pesticide applicators as well (even if unintentionally). The predominance of the pesticide paradigm and industry protection within regulatory culture therefore reinforce the narrative that pesticide drift is an unfortunate but ultimately small problem.

The relationship between narratives and their material and cultural context is an iterative, or mutually supportive, one: the material factors that strengthen the narrative are also, in turn, obscured or otherwise strengthened by it. The accident storyline further legitimizes a minimal regulatory response by obscuring the facts that the rules are vague, the rules are not always broken, reported illnesses represent the tip of the iceberg, and regulators sympathize with industry over other actors. In short, the dominant narrative justifies and naturalizes the minimal regulatory response to pesticide drift.

Justice in the Work of the Regulatory State

Regulatory failure in the case of pesticide drift is not simply a consequence of industry influence, malfeasance, broken scientific tools, and misleading stories that make it all seem reasonable. In addition to those causal factors, many of the pesticide regulatory and legislative reforms that compromised regulatory agencies' attention to pesticide drift were the outcome of predominant ideas of socially just relations between the state and society.

Like the broader environmental regulatory state, the pesticide regulatory apparatus reflects a *utilitarian* rationality: state interventions in environmental affairs should produce the greatest good for the greatest number. This is evidenced most clearly in the fact that FIFRA, the statutory foundation for pesticide regulation at the federal level, requires all pesticide regulatory decisions to be based in a cost-benefit analysis. Although reasonable in theory, cost-benefit analysis proves to be an inadequate basis for addressing environmental inequalities. Like utilitarian logic in general, cost-benefit analysis is blind to questions of distribution and equality. Also, many costs and benefits cannot be quantified, many institutional and social barriers prevent various stakeholders from participating in decisions about costs and benefits, and the process does nothing to address the underlying causes of those barriers.

Since the 1970s, a *libertarian* vision of justice has directly and dramatically shaped regulatory reform throughout the world in problematic ways. In the past few decades, libertarian political philosophy has taken on a life of its own. Neoclassical economists like Friedrich Hayek and Milton Friedman applied libertarian ideas to the realm of economic, social, and environmental policy analysis, and all major leaders in the Western world have stridently put those libertarian-inspired recommendations into practice. The libertarian conception of justice is the ideological heart and soul of the "neoliberalization" of environmental politics of the past thirty years: the reductions of government expenditures for environmental programs, elimination of regulations and other so-called barriers to trade, devolution of regulatory responsibility to the local level, and shift by the state, industry, and many activists alike toward voluntary and market-based solutions to environmental problems.[89] As Helga Leitner and her colleagues have stated, neoliberal reforms stemmed directly from the belief that "markets are the best, most efficient, and socially optimal means of allocating scarce resources in virtually all realms of life."[90] In 2001, conservative political strategist Grover Norquist memorably exemplified the libertarian sentiment when he declared on National Public Radio, "I don't want to abolish government. I simply want to reduce it to the size where I can drag it into the bathroom and drown it in the bathtub."[91]

The spread of neoliberal reforms, institutions, and dispositions comes in part from the selective, incomplete hailing of libertarian notions of justice, particularly by political leaders (most notoriously Ronald Reagan and Margaret Thatcher, but continuing up through the present day) who promote libertarian-inspired reforms that align with a neoconservative social agenda calling for greater "individual responsibility" to solve social problems. Libertarianism also has gained traction through the growing critiques of the perceived inefficiencies and inefficacies of Keynesian-era centralized regulatory structures—critiques that are not necessarily motivated by a libertarian ideal but whose lambasting of the state nonetheless provides a context that nurtures libertarian ideas and recommendations. Western countries' reactions to the spread of global socialism, and the decline of the civil rights and organized labor movements, similarly facilitate the rise of theories and policies that are blind to structures of oppression. Though neoliberal-style reforms may not always be conducted intentionally to pursue a libertarian vision of justice, the confluence of these (and other) factors supplied the conditions in which libertarian rhetoric could flourish as a fresh moral justification for revising the state's role in environmental politics. To be clear, libertarianism is by no means the only theory of justice shaping mainstream environmental politics today, and many libertarians surely would not approve of some of the purposes to which their ideals are applied (especially when libertarian rhetoric is used to legitimize repressive social policies). The libertarian conception of justice is nonetheless recognized as perhaps the most influential alternative to liberal egalitarianism in contemporary society; philosopher Colin Bird characterizes it as unmatched in its "contagion with intellectual circles, its (malign) influence on political discourse and public policy and its evangelical vigour."[92]

Fueled by a critique of centralized regulatory and welfare institutions, this increasingly prevailing preoccupation with minimizing the state's involvement in economic affairs translated directly into a downsized environmental regulatory state. In terms of pesticide regulation in the United States, this was manifested in legislators' cuts of environmental regulatory agencies' budgets, regulatory agencies' emphases on registering new "greener" pesticides and the concomitant glacial pace with evaluating the safety of old ones, and regulatory agencies' increasing reliance on market-based incentives and other voluntary agreements with industry. It is worth noting that both California DPR and U.S. EPA defend the success of their regulatory programs by gauging regulatory improvement in terms of the number of allegedly less toxic pesticides the agencies register for use, rather than the amount of dangerous pesticides (or the types of sloppy pesticide application methods) they actively restrict. For example, in their press release from 2006 about the most recent pesticide use data, California DPR officials boasted that the information suggested that a "dramatic increase occurred in the use of some newer, reduced-risk pesticides," and the DPR director exuberantly proclaimed, "This is just another indication that we are moving in the right direction"—despite the fact that such pesticides actually represent only a small percentage of the amount of pesticides used in the state.[93]

At the same time, pesticide regulation in recent years also reflects a *communitarian* vision of justice. While regulatory agencies have been slowly eviscerated and restricted, they also have increasingly created and facilitated mechanisms through which communities could come together to creatively and collaboratively address environmental problems. Pesticide regulators have created many opportunities for public comment at the local level, consistently stressed the value of devolved regulatory decision making, and facilitated "good neighbor agreements" between growers and residents as a means of solving conflicts through "common understandings." For instance, DPR's director recently lauded the Spray Safe applicator training program, declaring, "That kind of commitment you can't get with just the hammer, just the enforcement."[94] A CAC added, "It's far more effective if growers can kind of in a sense police themselves—It's kind of like a Neighborhood Watch program."[95] Promoters of regulatory devolution and other localization efforts argue that they comprise a new form of participatory democracy, and make politics more politically progressive. It is worth noting that communitarianism is quite compatible with libertarianism: these recent communitarian-inspired policies and institutions have gained traction in recent years by claiming to fill the gaps left by both the state's neoliberal retreat and the market-based mechanisms that replace environmental regulations.

Although communitarian narratives are persuasive and increasingly pervasive, many scholars have shown that a community's "shared understandings" often actually represent the interests of the politically powerful and sideline the opinions and experiences of those who have been marginalized. In contexts of deep social inequality and cultural oppression, the local is frequently as unrepresentative as broader scales of politics. Localist politics in some cases have been used to deliberately serve inegalitarian agendas, and in others they unwittingly end up doing so.[96] Ryan Holifield, for example, argues that the U.S. EPA's EJ programs, which draw on the communitarian principle of increasing opportunities to share experiences as a way to foster fairness, have been used to neuter political anger and direct activists to community participation mechanisms lacking any meaningful ability to shape local environments.[97]

One major problem is that libertarian and communitarian philosophers along with the policy advisers who adopt their principles (however intentionally and incompletely) grossly underestimate the extent of environmental externalities, inaccurately assume that the marketplace (for libertarians) or local politics (for communitarians) can provide the incentives needed for effective solutions to environmental problems, and accordingly dismiss the need for state intervention. The availability of new, less toxic pesticides has not translated to a decreased use of any of the most toxic and drift-prone pesticides (such as soil fumigants), and local politics are often more conservative than their higher-scale counterparts.[98] Yet such fallout produced by libertarian- and communitarian-inspired regulatory reforms is largely unrecognized, unmonitored, unquantifiable, or otherwise effectively dismissed. The fact that asymmetrical relations of power in agricultural communities make "negative externalities" like pesticide drift relatively invisible effectively legitimizes libertarian and communitarian regulatory reforms; since problems like pesticide drift are relatively invisible, official statistics misleadingly suggest environmental success. Moreover, the principles of justice that explicitly engendered regulatory reforms since the 1970s lend a veneer of credence as well as righteousness to a regulatory apparatus governed by the pesticide paradigm and industry interests.

Understanding regulatory work as partially guided by justice (however problematically) illustrates that pesticide drift arises in large part from increasingly popular ideas about the right way to solve environmental problems. As we will see, pesticide drift activists have a different set of ideas about what social justice means—and thus what environmental regulation should look like.

Pesticide Drift Activism: Response to Regulatory Failure

We were told with a straight face that it would be safe for our children to run around, ride their bikes, just feet away from people who were dressed virtually like astronauts as they fumigated the land.[99]

When "accidents" like this keep happening, it is no longer an accident but a poorly defined system that guarantees that such poisonings will continue.[100]

Although pesticide regulations and institutions are far from reaching their full health-protective potential, the vestige of a progressive regulatory apparatus means that pesticide drift activists do not have to reinvent the regulatory wheel. Activists' pursuit of regulatory change stems from an understanding of pesticide drift that differs radically from what guides regulatory agencies. Namely, pesticide drift activists are united by a shared belief that pesticide drift is a remarkably common—indeed, everyday— phenomenon, and they deliberately frame the problem in that way.[101] Many of the agricultural community residents I interviewed became politically active in response to a large-scale pesticide drift event—such as the Earlimart, Ventura, Arvin, and Lamont incidents described at the beginning of this book. Nevertheless, all these "resident-activists" emphasize that the large-scale pesticide drift incidents are really the outliers in a landscape of everyday contamination. In interviews with me and in community meetings that I observed, resident-activists recounted innumerable stories of suspected pesticide exposure—an unending narrative of pesticide drift events that they themselves, their family members, and their friends experienced at work, in their neighborhoods, and at school ("These things happen every day"). In addition to specific stories of suspected pesticide exposure, they described living in a literal mist of pesticides ("You can see the fog come out of the field when they are spraying"). Though some resident-activists and other agricultural community residents describe their experience of pesticide drift exposure as a dramatic, singular event, most emphasized how pesticide drift feels like a pervasive and all-encompassing presence in their everyday lives:

You can smell it. You can see it. When you drive, it gets on your windshield.

I figure I'm being drifted on by the entire Oxnard plain. And it's not target specific, and it's ongoing, and it lasts for months and it's in the carpet and it's in garden soil.

It's just become something normal for everyone here in the Central Valley. We just assume that we're going to walk out one day and get sprayed, or the tractor is going to be there.[102]

Many activists commented in interviews with me that pesticide drift is seen as a normal part of everyday life. As one noted, "In Alpaugh, when I lived there, I'd always hear stories about people getting sprayed, and losing their gardens, and it was like they'd just accepted it." Another remarked that new scientific research demonstrating high levels of certain pesticides in her neighborhood failed to shock many of her neighbors for precisely this reason: "It's like telling people that it's windy. We are so used to it." After describing to me an incident from 2009 in which several children were sprayed with pesticides while waiting for a school bus, another resident-activist stated that when the authorities "asked them if they were OK, they said, yes, they were fine. They were *drenched*, and they said they were fine. . . . People are just in that mode where we just accept the drift. . . . It happens all the time." This concern for drift as a routine phenomenon compels activists to demand regulatory interventions that address the long-term consequences of low-level exposures (i.e., not only the immediate experience of dramatic drift events). As one resident-activist stated,

It seems like the fumigants only make the news when there's some kind of acute, emergency exposure. . . . But for my neighbors and family, I'm more concerned with chronic and subchronic exposure and health problems that will manifest themselves years from now.[103]

Resident-activists emphasize that pesticides move because they and the environments in which they are used are inherently unruly, and hence difficult to control—winds change, temperatures shift, animals enter fields and break protective tarps, and people play, work, live, and travel near agricultural pesticide applications in innumerable unforeseeable ways. Activists thus cast pesticide drift in a new light: as a pervasive, under-regulated, everyday problem contributing to serious human illness and suffering. Drift is a part of daily life in agricultural communities, altering bodies over the long term in countless unquantifiable ways.

Many pesticide drift activists are politically motivated through their observations of illness in their communities, which they attribute to pesticides and other environmental contaminants. Many noted in interviews with me that local children are covered in persistent rashes, struggle with asthma, and play in pesticide-contaminated soil. Participants roll through lists of relatives, neighbors, and friends with health problems they attribute to pesticide exposure ("We're sick all of the time. Every year it is the same. . . . Everyone is sick every year during cotton-spraying season"). Resident-activists point to the debilitating, chronic health problems they attribute to pesticide exposure: asthma, other respiratory illnesses, heart attacks, cancers, migraine headaches, miscarriages, rashes, learning disabilities, behavioral changes, emotional distress, and others. According to one community organizer, Earlimart residents refer to their health in terms of "before the incident" and "after the incident." This stress on lingering illness broadens the impact of pesticide drift beyond the regulatory agencies' narrow frame of symptoms experienced at the time of exposure and also humanizes pesticide drift victims who are otherwise typically discussed as numeric trends. As one a group of leading public interest organizations active in pesticide drift debates noted in a technical comment letter to the U.S. EPA, "We would like to emphasize that poisoning incidents are not just a statistic to be viewed casually. These statistics represent real people's lives that have been significantly changed for the worse, often permanently."[104]

Their observations of pesticide drift and widespread illness provide many resident-activists with a sense of expertise and authority over the issue that occasional outside observers simply cannot get, sentiments they highlight in their statements at community meetings and in confidential interviews with me:

Being in the bull ring is not the same as watching from the outside.

You don't know what's out there. I know what's out there. I worked there forty-five years.

We need to educate the experts. We *are* the experts! . . . We know what we're feeling and what we're going through.

Though their suspicions and allegations about specific exposures and illnesses are not always verifiable or confirmed, their experience with everyday life in agricultural communities provides illuminating insights into how we understand pesticide drift.

In particular, pesticide drift activists reveal how social issues that pervade agricultural communities—poverty, job insecurity, systemic dismissal and disregard by authorities, language barriers, lack of legal status, and lack of access to health care and legal assistance—prevent adequate reporting, medical treatment, and analysis of pesticide drift, not to mention adequate regulatory responses to it. Playing a major role in pesticide drift activism, grassroots resident-activists regularly recount their stories along with their anger in oral and written form at formal regulatory hearings as well as in activist publications, press statements, and interviews with researchers. The following statements exemplify the ways that pesticide drift activists portray how various social issues render pesticide illnesses invisible, lead to official statistics that paint pesticide drift as a relatively minimal and unremarkable problem, and legitimize regulatory agencies' limited response.

Poverty inhibits many residents' abilities to escape pesticide exposure at home, as one farmworker with two young children explained:

We live in a trailer park in a mobile home, surrounded by orange groves. At night during the spraying season we get sick, with symptoms that wake us up. We have asked the owner of the trailer park to do something about this problem, at least let us know when spraying is going to happen, but the owner always tells us we can leave if we don't like the situation, that we should look for another place to live. But we can't afford to rent a lot in another park.[105]

At the same time, poverty and high competition for agricultural jobs also exacerbate pesticide exposure in the workplace. Farmworkers say that they feel pressured to accept pesticide exposure risks in order to keep their job. Several resident-activists with prior experience as farmworkers told me that when they asked for personal protective equipment to shield themselves from pesticide residues, their crew leader would say, "You want to protect yourself? Stay at home or get another job."

The fear of being fired also prevents workers from reporting suspected pesticide exposures. One community organizer with many years of experience as a field worker repeatedly told me that workers are convinced that those who report pesticide exposure would not only get fired but also get blacklisted ("put on the famous *lista negra*"), and thus be unable to get any job in that area (with employers saying to each other, "Don't hire him—he's a troublemaker"). One other community organizer similarly highlighted this point: "I talked to community people who have been exposed to pesticides. When we told them to go to the doctor, report to your supervisors, they said, 'I don't want to lose my job.'"

Poverty exacerbates the likelihood of pesticide exposure and prevents people from reporting suspected exposures in other ways as well. In interviews with me, farmworking women commented that health problems are a major hardship for them. Families can buy food or they can send someone to the doctor—not both. In cases of suspected pesticide exposure, many people resort to home remedies because they cannot afford to go to the clinic. Many women expressed considerable frustration with a system that requires those who suspect they have been exposed to pesticide drift to pay for their own tests in order to establish proof, arguing that those people do not earn enough money to go to the doctor in the first place and should not have to bear the responsibility for someone else's wayward pesticides. The fact that poverty and other forms of marginality inhibit people's abilities to report pesticide exposure came up frequently in my interviews, such as in the following observations:

The nearest hospital is in Delano, which is seven miles away. And then the nearest hospital north is Porterville or Tulare. A lot of people don't have transportation to get to any of these. Not even seven miles away. That's a big deal for a lot of people.

They said they need sufficient proof. How do you prove it with no money? No money, no doctor!

Regulations don't fit the reality. Regulations assume that people have laundry, speak the language, have technical knowledge, information, and money.

Additionally, as numerous farmworking community residents explained to me, the fact that many immigrants lack legal status also deters many people from reporting pesticide exposure:

There's incidents that are happening all of the time, they just go on not reported. Sometimes the people don't want to get involved—fear of retaliation—or they just simply don't know what it is. Especially the undocumented. You know, "We're not saying anything. We'll get deported. We'll get fired. Our families back home depend on us working." [So they are] afraid to speak out and have the fear that they're going to be deported or going to be fired.

Farmworkers generally just want to work, eat, raise their kids, pay their bills, and not make any noise or cause any problems. They don't want attention, so they don't speak out.

For people without legal status, why would they risk everything to voice their opinion?

Here, where a lot of jobs rely on ag and people think that their immigration status would come into play there and that maybe they would get the immigration [authorities] called on them, they won't report.

One big issue is that people are afraid to report or speak up because of legal status. People are afraid.

We don't want to cause any problems. We don't want to be deported.

One community activist I interviewed noted that community members are afraid to report suspected pesticide drift incidents because they fear retaliation. Pesticide applicators have reportedly deliberately doused houses with pesticides after they heard that residents there complained to the authorities of a suspected drift incident: "People are scared. They don't want to make a fuss. They don't want retaliation."

Some residents do report pesticide exposure, yet many who do feel utterly dismissed as well as belittled by their employers, medical professionals, and regulatory officials. For example, one politically active resident recalled an event in which a neighbor of hers in a small Tulare County town testified at a council meeting that she had been recently sprayed by pesticides, and as a result, had developed a rash.

So all of these farmers showed up, and I happened to be at this meeting, and god they were just mean to her, they were brutal, they called her crazy. I was amazed how she was treated and dismissed and accused of being nuts, and I had no reason to doubt what she had said.

Another former farmworker reported that field supervisors often respond to workers who express concern about a possible pesticide exposure by asserting, "You're just drunk from the weekend and that's the reason. It's not because of pesticides." Or for women, "It's because you are pregnant. It's not because of pesticides."[106]

People involved in pesticide drift incidents report having been ignored and disbelieved by local emergency response crews and health clinic staff members, who typically have little or no training in recognizing and reporting the symptoms of possible pesticide exposure.[107] The consequences of such dismissal became painfully clear in Lamont in October 2003, where officials' failure to take seriously the residents' pesticide illnesses one night enabled the pesticide applicator to continue with the second half of the field fumigation the following day—sickening over two hundred additional people with chloropicrin.[108] One resident-activist described for me her friend's experience of this particular drift event in the following way:

Emergency crews came into their house, where my friend and her husband were having difficulty breathing, where the children were on the floor throwing up. And the crew said, "Nothing's wrong here, we don't smell anything." The ambulance was across the street, and she said she didn't know if she was going to live or die. No one helped them. She said that their neighbor chased the ambulance, but they laughed at him and drove away.

At a DPR hearing, one young farmworker described how she had been exposed to an unknown pesticide while working in fields while pregnant. Two days later, she had a miscarriage. Because the pregnancy until that point was fine, she felt that the pesticides might have been a likely cause of the miscarriage. She said that the doctors refused to believe her assertion that she had been exposed to pesticides, and did not test her for pesticide exposure or report the possibility of it.

One community organizer with over twenty years of experience working in the fields explained how the consistent dismissals by supervisors and doctors has a chilling effect on people's willingness as well as ability to report pesticide exposures. She noted that when farmworkers report suspected pesticide exposures to their supervisor, the supervisor often dismisses them by responding,

"What were you eating last night?" or "You've been drinking," or "You're pregnant," or "Probably you have the flu," and "Well, how do you know that you got sick from the pesticides?" or "The chemical we are spraying is not dangerous. It's soap! It's not dangerous!" Anyway, the worker probably decides, "OK, I'm going to check with the doctor." And the doctor says, "No. It's not pesticides." The company says the same, the doctor says the same. What option do you have? If I'm the boss, and I say to you, "No, it's not pesticides," and then you [get the same response from] the doctor, you don't have another option! . . . The two main people—the boss is the one who has to give you the pass for going to the doctor, and the doctor is the only person you are going to trust—they say, "No."

Pesticide drift activists also commented that regulatory officials typically fail to take seriously residents' reports of pesticide drift. One activist asserted that delays in regulatory response compromise the ability to determine whether drift occurred:

Even if you're lucky enough to get someone [from a regulatory agency] to come out—and they don't, usually—unless they get there in a half an hour to an hour, the levels are usually much lower, or they're undetectable. Then they say, "There's nothing here now, so probably there was nothing here before." Where's the scientific method in that? It's just pure conjecture.[109]

Pesticide drift activists explain that a slow, inadequate, or disrespectful regulatory response to residents' reports of pesticide drift creates a negative feedback of suppressing reports of other pesticide exposures:

If people see that the officials do not do anything when other people report pesticide exposure, then they say, "Why should I report this? Nothing will be done."

In areas where pesticide drift is frequent, many residents have stopped reporting drift incidents to regulators because of the apparent futility in obtaining a response that would result in relief.[110]

In sum, resident-activists' stories demonstrate that a range of social factors inhibit the likelihood that any pesticide exposure report will make it into the official statistics of the extent of pesticide drift: victims do not know how to report suspected pesticide exposures, do not want to jeopardize their employment, cannot afford to attract the attention of the law because they (or a family member) lack legal status, do not have access to medical care, and/or have a difficult time effectively communicating with local officials or other investigators. These social constraints work in concert with a range of more technical factors (e.g., many acute symptoms of pesticide exposure mimic common ailments, and thus seem unremarkable to victims and their doctors, and some pesticides do not cause any acute symptoms at all), together obscuring most pesticide drift and illness from public view and official statistics. Of course, it must be recognized that residents' suspicions of pesticide exposure are not always accurate; sometimes the drift they see is just water or seeds, or the symptoms they experience are due to other factors. The discussion above, however, illustrates that agricultural environments are saturated with toxic chemicals, and suspected pesticide exposures are often either ignored or impossible to definitively ascertain. Given the rarity of air-monitoring studies, the underreporting and inadequate investigation of suspected pesticide exposures is tremendously consequential: without evidence of pesticide exposure and illness, the problem disappears from view. In other words, without an apparent hazard, the status quo looks just fine.

The combination of personal observation along with a sense of abandonment is discouraging and disempowering for some residents, but it gives others the confidence and conviction needed to become politically active, publicly share their impressions of a contaminated landscape, and offer explanations that differ from the dominant accident narrative. Resident-activists use these personal accounts to show that the phenomenon of pesticide drift is pervasive, explain why it is largely invisible, and accordingly argue for more health-protective regulatory restrictions on agricultural pesticide use. While some make blanket statements about wanting to ban all pesticides, most pesticide drift activists I have met proclaim much more targeted regulatory reform goals that pivot around greater regulatory restrictions on the use of the most drift-prone and highly toxic pesticides.

In their quest for strengthened pesticide regulations, many pesticide drift activists have joined a growing chorus of activists and scholars who express concern about the ability of experts to accurately predict the risks posed by pesticides.[111] Like their counterparts focused on other environmental issues, pesticide drift activists draw attention to the factors that compromise experts' understandings of issues like pesticide drift. In technical comment letters and spoken testimony to the U.S. EPA and DPR regarding specific pesticide regulatory decisions, scientists participating in pesticide drift activism point out the specific gaps in scientific knowledge that compromise the predictive capacity of risk assessments (such as the gaps in research on certain chemicals' abilities to cause various health effects, and the universal lack of knowledge about the effects from exposure to the pesticide combinations to which people are really exposed in agricultural communities).[112] Similarly, resident-activists' testimony demonstrates how a litany of factors compromise scientists' abilities to understand, track, and predict pesticide exposures and illnesses: the ways that poverty, job insecurity, language barriers, and legal status both exacerbate pesticide exposures and obscure them from official pesticide illness data sets. To ameliorate these various limitations of risk assessment and the regulations they produce, pesticide drift activists call for a more health-protective model of chemical regulation that can proactively anticipate as well as account for such data gaps, realistic exposure scenarios, and social issues.

Like many other environmental activist groups throughout the world, pesticide drift activists argue that regulatory agencies can do exactly this—namely, by adopting what has become known as *the precautionary principle.*[113] PAN defines the precautionary principle as

the concept that we must act to reduce potential as well as proven hazards, even in the face of uncertainty about the extent of these hazards. Among other things, this means taking action to eliminate exposures to potentially damaging substances. In our work, this principle is even more obvious, because so many ecologically sound, healthy alternatives to pesticides exist.[114]

Thornton explains the precautionary principle and its relationship to standard scientific practices in the following way:

Toxicology, epidemiology, and ecology provide important clues about nature but can never completely predict or diagnose the impacts of individual chemicals on natural systems. The implications for policy are obvious: since science leaves so much unknown, we cannot afford to make risky bets on its predictions or wait to protect health and the environment until we know for certain that some substance or technological practice has caused injury. Instead, we should avoid practices that have the potential to cause severe damage, even in the absence of scientific proof of harm. This rule, called the precautionary principle, is common sense: and says that we should err on the side of caution when the potential impacts of a mistake are serious, widespread, irreversible, and incompletely understood, as they are with the hazards of global toxic contamination.[115]

Raffensperger and Tickner describe the history of the precautionary principle. Recognizing that some environmentalists in addition to workplace health and safety experts have long advocated the concept of precaution, and that the concept gained significant traction with the publication

of Carson's *Silent Spring* in 1962, they detail how the term precautionary principle was formalized in the Wingspread Statement of 1998 at a
conference in Racine, Wisconsin.[116] Sociologist Phil Brown explains that
activist groups deliberately disseminated and advocated the precautionary principle following the Wingspread conference and that these efforts
helped gained legitimacy for the concept within activist, academic, and
policy circles by the end of the twentieth century.[117]

The precautionary principle stems from a critique of risk assessment but
is not a blanket rejection of it. Instead, as Kerry Whiteside emphasizes in
his recent book, *Precautionary Politics*, the precautionary principle must
be understood as an appropriate guiding logic to help make decisions in
cases where risks are poorly understood and where scientific disagreement
exists: "The precautionary principle is not meant to replace science-based
risk assessment across the board but to indicate that risk assessment does
have its limits—and when these limits are reached, special obligations en-
sue."[118] Those special obligations include taking seriously the gaps in scientific knowledge about the potency of many chemicals, and as pesticide drift
and other activists demonstrate, recognizing the ways that social issues constrain experts' abilities to assess the extent of chemical exposures. Although
Whiteside therefore conditionally accepts risk assessment and cautions his
readers against misinterpreting the precautionary principle as a wholesale
rejection of it, he also makes it clear that the precautionary principle works
with a different underlying principle: decision making within the framework
of the precautionary principle "takes place under a general guideline favor-
ing prevention and the search for alternatives to environment-damaging
practices rather than arguing for an 'acceptable level' of risk."[119]

Accordingly, proponents of precaution advocate a variety of policies
and regulatory changes that both integrate precaution into current risk-
assessment-based regulatory practice and use precaution to restructure risk
assessment more fundamentally toward pollution prevention. To integrate
precaution into the existing model of risk assessment, scholars and activists
recommend building larger uncertainty factors and delaying a chemical's
registration when key scientific data are missing, rather than misinterpreting a lack of data as a lack of threat; defaulting to a health-protective option
where the science is unclear; studying chemicals as classes, making reasonable and health-protective guesses about chemicals that share structural
characteristics, and regulating them based on that; increasing funding for
environmental monitoring, toxicology, and epidemiology; reinforcing the
independence of regulatory bodies from industry influence; prohibiting a
technology if it is simply too poorly understood and/or too dangerous; and

reorienting environmental regulation toward pollution reduction instead of pollution management.

Proponents of the precautionary principle also call for substantially increasing opportunities for public participation, in order to bring into the regulatory process the relevant insights and perspectives that have historically been marginalized from it and that are relevant to regulatory decision making. Additionally, greater public participation is needed to open up for democratic deliberation the value judgments that always take place within scientific practice. In their critical evaluation of risk assessment, Raffensperger and Tickner argue that its seclusion from democratic debate "allows agencies to justify and defend their decisions to the courts, businesses, and the public in the guise of objective, unbiased numbers, avoiding mention of the values implicit in decisions affecting public and environmental health."[120] The precautionary principle, in contrast, calls for increased public participation, which can open various subjective questions for public scrutiny and reconsideration: Which costs and benefits should be considered, and how should they be calculated and compared? How large of safety margins can adequately account for various forms of scientific uncertainty? What constitutes an "acceptable" risk of cancer? Who gets to decide? How realistic and reasonable are various risk mitigation measures? Echoing Paigen's quote at the start of this chapter: Do the judgments presently embedded in standard scientific protocol seem reasonable and fair? Two pesticide drift activists highlighted this conviction in their letter to the editor of a local paper in 2007: "The question of what is safe enough for children and others is a question of both science and policy. It is rightfully answered by society as a whole, and in particular by those at risk of exposure."[121] Brown and numerous other advocates of community-based participatory research push this call for participation even further, advocating for increased public participation in scientific research itself, both to improve its accuracy and further democratize the decisions that affect people's lives.[122]

The precautionary principle also suggests the need for a more fundamental restructuring of risk assessment itself. First, to address the problems inherent in risk assessment's reductionist analysis of singular chemicals, the precautionary principle indicates the need for *cumulative risk assessment*. In contrast to risk assessment's methodology of looking at chemicals "in isolation, in laboratory conditions, and truncated from social factors," cumulative risk assessment is needed to identify the ways that chemicals interact with each other and other environmental conditions.[123] Second, to address the fact that risk assessment's myopic focus on singular chemicals cannot account for whether any given chemical is beneficial in light of the alternative options—that is, whether introducing that chemical's risk into the environment (or continuing to allow it) is even necessary to begin with—proponents of precaution advocate *alternatives assessment*.[124] Alternatives assessment shows how a given pesticide stacks up against other pest management options, in terms of both risks and benefits to a broad range of actors.

Pesticide drift activists' vision of a socially just regulatory framework is one that puts the precautionary principle into practice. They argue for building precaution into pesticide regulation to acknowledge that standard scientific practice is unable to unravel the tangle of causes and effects when chemicals work in additive and synergistic ways, does not address the total pesticide burden to which agricultural community residents are exposed, and cannot account for a tremendous range of social issues that worsen and obscure pesticide drift and illnesses. Pesticide drift activists' precaution-based demands include the following:

• incorporate health-protective uncertainty factors and other safety margins into pesticide risk assessments whenever the relevant science is missing or uncertain

• subject highly toxic pesticides to robust alternatives assessments

• phase out chemicals for which less toxic alternatives exist

• phase out the use of dangerous application methods for highly toxic pesticides

• standardize pesticide regulations across all counties

• implement formal and strong public health agency oversight of pesticide regulatory decisions (including, though not limited to, appointing public health experts to lead pesticide regulatory programs)[125]

• strengthen policy support for alternative pest management research and outreach along with regulatory protections of market-based measures that support farmers who practice less toxic agriculture (such as the organic label)

• increase opportunities for public participation such that people are actually able to influence the decisions that affect their lives

Pesticide drift activists also seek to simply clean up current regulatory practices. As such, they call for regulatory agencies to

• conduct robust risk assessments and use those—instead of adulterated versions of them—to guide risk management decisions

- reduce industry influence over science and regulatory decisions
- enforce existing laws (e.g., prosecute violators, for instance, and respond to all reports of drift and other concerns about pesticide exposure quickly and respectfully)
- design punishments that compel adherence to existing laws
- strengthen cost-benefit analyses to account for all costs of a chemical (e.g., the costs to health, plus the costs of research and other regulatory practices needed to research, monitor, and regulate its use)

In contrast to industry actors' disparagements of pesticide regulations as inherently regressive and debilitating, pesticide drift activists frame health-protective, precaution-based pesticide regulatory restrictions as *progressive*—as an essential component of making California agriculture more innovative and environmentally sustainable. An activist-scientist involved in political debates over pesticide drift, for example, summarized her critical analysis of the fumigant metam sodium by arguing for "a precautionary approach": "aggressive alternatives assessments that prioritize and advance truly safe and sustainable alternatives to hazardous pesticides."[126] To substantiate this contention, activists point out the availability of alternative pest management systems and evidence that regulations can prompt their further development. For instance, pesticide activists often point out that the FQPA of 1996 is widely recognized by agricultural scientists as spurring the development of alternative, greener technologies.[127]

Grassroots activists and their scientist colleagues pursue these regulatory reform goals through a wide variety of practices, including collecting many different kinds of data, staging protests, publishing reports for a wide variety of audiences, educating and organizing residents of agricultural communities, preparing and delivering oral testimony at formal regulatory hearings, joining agencies' public advisory committees, organizing "toxic tours" to educate decision makers, submitting written comments on proposed regulatory reforms, and holding press conferences (see figure 4.2).

Activists encourage community residents to share their own stories of pesticide exposure and help residents learn how to navigate the regulatory system. In addition to holding community educational meetings, CPR created a "Community Response Guide" to help educate residents about how to respond to pesticide incidents and how to become more politically active.[128] Some resident-activists use cameras and videos to try to document pesticide drift and make pesticide applicators more accountable for their

Figure 4.2
Pesticide drift activist Teresa DeAnda at a news conference in Arvin, California, on February 9, 2010. At the event, the activists announced Kern County's new buffer zone restrictions on applications of restricted material pesticides near schools. *Source*: Photo by David Chatfield.

actions. For example, Teresa DeAnda films every pesticide application that takes place across the street from her house, and Ignacio Azpitarte (see figure 4.3) has regularly filmed pesticide applications ever since he was hospitalized after exposure to pesticide drift.

In California, the regional and umbrella organizations most involved in pesticide drift activism—CPR, PAN, the Center on Race, Poverty, and the Environment, and California Rural Legal Assistance Foundation—have deliberately devoted considerable attention to the southern end of the Central Valley because of the severity of pesticide drift and relative lack of activist resources there. These NGOs work with local community-based organizations to educate residents about the health effects of pesticides, residents' rights to not be exposed, and mechanisms for reporting pesticide drift and exposure. Community groups in Tulare, Kern, and Stanislaus counties have successfully pressed their CACs to establish new county-level, health-protective buffer zones around schools, residential areas, and other sensitive sites for many applications of restricted use pesticides.[129]

Figure 4.3
Ignacio Azpitarte sits in his home in July with a video of a crop-dusting helicopter spraying near his house in Weedpatch, near Lamont, California. Azpitarte has a collection of still and video cameras to document overspraying. *Source*: Photo by Dan Ocampo, *Bakersfield Californian*, 2005.

Pesticide drift activists also help to design and lobby for various state-level legislation, such as the Pesticide Exposure Response Act (SB 391, passed in 2004), which mandated improved emergency response procedures to pesticide drift incidents and reimbursement of uninsured victims' medical bills. As another example, the Northwest Coalition for Alternatives to Pesticides played a major role in pushing a bill through the Oregon state legislature in 1999 that mandated the Oregon Department of Agriculture to develop and implement a pesticide use reporting system.[130] NGOs also push their pesticide drift campaigns to the global level. They have been key, for instance, in designing and sustaining international agreements such as the Stockholm Convention, a treaty to eliminate toxic, persistent, bioaccumulative chemicals that are capable of long-range transport.[131]

Pesticide drift activists also periodically file lawsuits against regulatory agencies to force them to fulfill their mandates. For example, as part of their broader effort to emphasize the everyday nature of pesticide drift and press for state-level regulatory reform, activists in California have sued the state for violating the Clean Air Act. These lawsuits specifically target the state's failure to control pesticides' contributions to ground-level ozone (a central component of smog) and have compelled the California EPA and regional air districts to start taking responsibility for this aspect of pesticide pollution.[132] As a further effort to force pesticide regulators to recognize and reduce airborne pesticides, CPR has bought a lawsuit against DPR for failing to implement the TAC act.

In addition to formal lawsuits, pesticide drift activists periodically formally petition regulatory agencies to make health-protective regulatory changes. For example, in October 2009, PAN, the UFW Union, the California Rural Legal Assistance Foundation, and other organizations petitioned the U.S. EPA to better protect children in rural communities by mandating buffer zones around sensitive sites like schools.[133] Furthermore, pesticide activists in California have petitioned several times to have the state's pesticide risk assessment formally moved from DPR to the OEHHA, which conducts all of the other chemical risk assessments for the state and has a considerably more health-protective track record than does DPR.[134] Activists argue that DPR's charge to conduct both risk assessment and risk management constitutes a conflict of interest, which they allege is compromising DPR's ability to implement health-protective pesticide regulations.[135]

To further combat industry actors' disproportionate influence over regulatory decisions, activists protest regulators' appointments of industry leaders to key positions in regulatory agencies and critically publicize industry requests for special treatment from regulatory agencies. In other campaign work, PAN explicitly challenges how such appointments and industry's extraordinary lobbying resources enable it to have disproportionate influence over regulatory decisions.[136]

Grassroots pesticide drift activists work together with scientists and other researchers, who test residents' claims and are often able to substantiate them with other forms of data that gain traction in the current regulatory system that privileges scientific discourse and data. As further evidence that pesticides are unruly and that this uncontrollability poses grave risks to public health, scientist-activists point to studies that show the ambient concentrations of some pesticides to exceed health benchmarks even when applicators were clearly on their best behavior.[137] Pesticide drift activists also have started to conduct their own scientific monitoring studies and publish reports that compare the ambient concentrations detected against health benchmarks. In these "Drift Catcher" projects, community residents work with local and regional NGO staff members and PAN scientists to sample air for pesticides, and then use the results to press for regulatory reform.[138] Figure 4.4 below illustrates how the Drift Catcher results are represented.

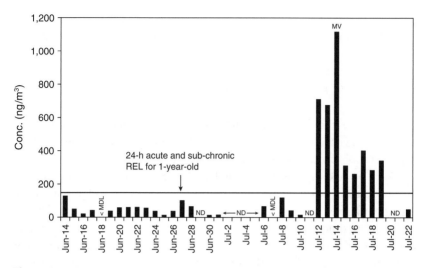

Figure 4.4
Representation of one Drift Catcher project results. *Notes:* "REL" indicates the Reference Exposure Level, a health benchmark calculated from the U.S. EPA's acceptable daily (twenty-four-hour) dose for acute and subchronic exposures. "ND" indicates no data. "MV" stands for minimum value and indicates that the value for that particular sample may actually have been higher. "< MDL" indicates that the amount was below the method detection limit. *Source:* PAN 2006, 20.

PAN has also conducted several "Biodrift" projects in which participants, in addition to sampling the air outside their homes for evidence of drift, sample their own urine and have it tested for evidence of pesticide exposure (a practice called "biomonitoring").[139] Sampling their own air or bodies for pesticides strikes many pesticide drift activists as an exciting way to validate their suspicions, especially since they see themselves as guinea pigs living squarely in the middle of scientific practice simply by virtue of residing in agricultural communities. Drift Catcher projects have been conducted in fifteen locations in nine U.S. states, and participants represent a wide range of demographic groups: Latino farmworkers (Washington, Florida, and several California sites), middle-class high school students (Florida), Native Americans (Minnesota), and other residents of agricultural communities (Indiana, Maine, North Carolina, Minnesota, Colorado, Oregon, and Hawaii). Such efforts are going global; PAN has also trained activists in South Africa, Germany, the Philippines, and Senegal to use the Drift Catcher.

PAN and local organizations have used the Drift Catcher data to raise public awareness in the local participants' communities about pesticide drift through community meetings, press releases, television and newspaper coverage, and letters to the editor of local newspapers. Local participants have drawn on their data to strengthen their testimony in legal, legislative, and regulatory hearings about the human impact of pesticide exposure. The Farm Worker Pesticide Project in Washington State used its Drift Catcher results to successfully lobby the state legislature to secure funds for a Washington State air-monitoring program for pesticides. PAN and many local organizations have used Drift Catcher data in comment letters and other technical documents as well to influence state and federal reregistration decisions for specific pesticides.

While providing new Drift Catcher data in those technical comment letters, scientist-activists use the opportunity to identify and publicly criticize the innumerable flaws in specific regulatory risk assessments. That is, at the same time that pesticide drift activists draw on and contribute to conventional scientific research through the Drift Catcher program, they continually point out that the scientific perspective is an imperfect one. Activists' stories about how social inequalities and various forms of oppression widely render pesticide exposures and illnesses invisible illustrate the limits of scientific understandings of pesticide drift. Like antitoxics activists in other contexts, pesticide drift activists both use and critique scientific knowledge. They work with a nuanced and expanded conception of expertise, strategically integrating scientific data with residents' personal experiences, observations, and assessments.

Justice in Pesticide Drift Activism

Pesticide drift activism provides a lens into the different notions of justice that shape environmental politics in two overarching ways. First, pesticide drift activists reject the communitarian- and libertarian-based approaches that increasingly govern the environmental arena, arguing that they are fundamentally unable to meaningfully address pesticide drift. Pesticide drift activists explicitly critique these recent trends in the regulatory arena. As one activist stated, "Voluntary programs are a euphemism for 'loophole.'" Another emphasized that pesticide drift activists push for regulatory restrictions precisely because industry-friendly, voluntary solutions simply have not redressed the problems associated with pesticide use:

What we're seeing with the increase [in levels of pesticide pollution] is that state agencies in the face of a serious health crisis are just adopting a business-as-usual approach. They've maintained using voluntary measures despite being sued and despite seeing these increases, and clearly more action needs to be taken.

Second, pesticide drift activists show why pesticide drift persists despite present regulatory and industry efforts, and in so doing, they exhibit a different notion of what justice means. Rejecting the libertarian and communitarian assumptions that unfettered markets and communities' mythical shared understandings can produce socially just outcomes, pesticide drift activists, like the broader EJ movement, assert that justice is a process of accounting for and combating inequality, oppression, a lack of meaningful participation, and inadequate basic capabilities. This alternate vision of justice is evident in pesticide drift activists' explanations of regulatory failure and recommendations for change. I discuss each of these four components of pesticide drift activists' vision of justice below.

Pesticide drift activists' critiques of regulatory failure highlight and problematize the striking *inequality* in pesticide exposure and regulatory response. Activists stress the disproportionate exposure experienced by people living in agricultural communities relative to the rest of the population, noting that environmental laws protect some social groups (e.g., consumers) much better than others (e.g., agricultural community residents). They also criticize the unevenness of pesticide regulations across space— the variations in local regulations that leave people in some places better protected from pesticide drift than people in other places. At the same time, activists point to as well as publicly protest the chemical and agricultural industries' disproportionate influence over legislative and regulatory decision-making processes relative to the public. Pesticide drift activists therefore advocate solutions that redress such inequalities. For example, to equalize pesticide regulations across space, pesticide drift activists tend to prioritize regulatory change at the state and federal levels.

Pesticide drift activists explain regulatory failure by drawing attention to the institutional and cultural processes that prevent many agricultural community residents from participating on an equal footing in daily life. These processes function together as *oppression*, and recognizing the various forms of oppression at work here helps to explain the persistence of pesticide drift. Following the lead of political philosopher Iris Young, the oppression I refer to is not (only) an artifact of tyranny or overt force but instead is largely structural: "the vast and deep injustices some groups suffer as a consequence of often unconscious assumptions and reactions of well-meaning people in ordinary interactions, media and cultural stereotypes, and structural features of bureaucratic hierarchies and market mechanisms—in short, the normal processes of everyday life."[140] In other words, as Nancy Fraser defines it, oppression refers to the state of "being prevented from participating as a peer in social life."[141] Pesticide drift activists highlight the social practices and institutions that structurally oppress so many residents of agricultural communities. Much of this work focuses on the oppression experienced so acutely by immigrant farmworkers and their families—political disenfranchisement, poverty, racism, and legal vulnerability. Immigrant farmworkers and others living in poverty, without legal status and/or with limited English language skills, are not only often disproportionately exposed to pesticides but also possess disproportionately less ability to leave polluted spaces, report pesticide exposure, and access medical and legal services in cases of exposure. These institutional and cultural processes both compromise regulatory efforts and make a considerable amount of pesticide exposure and illness invisible, thereby legitimizing an inadequate regulatory response.

Consequently, activists identify as members of social groups that experience oppression and marginality in ways that exacerbate pesticide exposure and call for regulators to recognize rather than ignore those relationships. This argument is captured in the following statement, in which a resident-activist from a predominantly Hispanic, low-income community in the Central Valley described in an interview with me a statement that she made to her County Board of Supervisors:

[The California EPA's statistical categories are] "white," "Pacific Islander," "African American," and then they have "other." So there is no Hispanic [category]. . . . I said, "We are home to the tire incinerator, the garbage burner, the Superfund site, and the landfill expansion. If you are going to be continuing to practice environmental racism on the west side [of the Central Valley], the least that you could do is recognize the fact that we are here. Call us something other than 'other.'"

Pesticide drift activists maintain that these forms of oppression should not only be recognized but also be accounted for within regulatory practice. Many pesticide drift activists explicitly reference the precautionary principle as a framework for doing just that. They call for regulatory officials to implement precaution-based restrictions on the most toxic and drift-prone pesticides to account for the ways in which oppressive social relations undermine the ability of current regulations to protect public health. They also demand the incorporation of health-protective, precautionary measures within risk assessment; implementation of regulatory reforms that are designed to protect even the most vulnerable bodies (rather than purportedly average ones), protect against worst-case incidents, and account for actual exposure scenarios (i.e., the cumulative and synergistic effects of exposure to chemical combinations); restrictions on industry influence over regulatory decision making; and aggressive enforcement of existing laws.

In order to start to redress the oppressive relations that exacerbate and obscure pesticide drift, many regional organizations involved in pesticide drift deliberately pursue county-level reforms in areas characterized by both high rates of pesticide use and marginalizing forms of social oppression. To further redress these various forms of oppression, many pesticide drift activists are politically active in immigration rights advocacy, workers rights activism, and health care reform. Many pesticide drift activists view their involvement in pesticide politics as just one more way to build power within historically marginalized communities and thus combat the oppressive social relations they experience.

Pesticide drift activism also highlights the processes through which the pesticide regulatory apparatus structurally oppresses people who do not adhere to the pesticide paradigm. Within regulatory agencies, the pesticide paradigm is perceived as normal and necessary, whereas people who question the safety and necessity of pesticides along with the adequacy of pesticide regulation are publicly disparaged and dismissed as unreasonable, irrational, emotional, and unrealistic. Pesticide drift activists emphasize that the ridicule and denigration of concerns about pesticide exposure intimidates other workers and residents into not reporting their own experiences. In this way, the pesticide paradigm functions as what Young calls "cultural imperialism."[142] Fed up with being dismissed or disparaged by industry elites and regulatory officials, pesticide drift activists question and denaturalize the pesticide paradigm. They contend that pesticide-intensive agriculture is neither inherently good, necessary, nor reasonable, and insist that other viewpoints be honored and respected. The precautionary principle offers a widely recognized framework through which pesticide drift activists can build alliances with other EJ and antitoxics activists fighting similar dominant paradigms in other contexts.

Pesticide drift activism also illuminates the injustice in the ways that oppressed groups have largely been excluded from *participating* in pesticide regulatory practice. Participatory parity—"equal, informed, respectful participation"—is a core principle of the EJ movement.[143] Pesticide drift activists join many formal opportunities to participate in the pesticide regulatory process. Yet they note that they feel effectively excluded from meetings and hearings that are held in locations that are difficult to travel to, held at times of day when people typically are at work, lacking Spanish language translation, or not widely advertised through mechanisms that reach community residents. Activists emphasize as well that participation requires actively and systematically seeking and incorporating marginalized groups' perspectives seriously throughout the pesticide regulatory process. In an interview with me, one activist leader pointed to the need for "meaningful participation by the people who are affected—from day one. From an EJ perspective, if DPR was curious about doing anything about a response to community poisonings, don't put a word on paper until you've at least had a few good talks with communities." Moreover, participation must be substantive and authoritative, not just symbolic. Activists want to help define the terms of debate rather than just to sit at the table. They want to redefine the questions that regulators pursue, such as including the following: Is this chemical even necessary? What other nonchemical alternatives exist? Without the ability to influence material outcomes, public participation opportunities represent empty communitarian rhetoric, a co-optation of activist concerns, and ultimately a drain of activists' time and energy, instead of a meaningful opportunity to effect regulatory change. (One activist described her participation in one of DPR's consensus-based "EJ" advisory groups as "business as usual with a better public face.")[144] Thus, activists also participate in the regulatory process through more traditional, oppositional mechanisms (such as lawsuits, testifying at hearings, and public demonstrations).[145]

Justice requires more democratic participation in the generation of scientific knowledge, too. Pesticide drift activists' Drift Catcher program serves many of the same functions as the community-based participatory research projects that other scholars have showcased in their studies of antitoxics and EJ activists. Such projects can highlight "research silences," facilitate new or better scientific methodologies, identify missing variables, shift research attention to the upstream environmental and political causes of illness, boost civic engagement, illuminate the value judgments on which scientific standards (e.g., statistical significance) are based, and help citizens influence environmental and health policy, pollution levels, and industry accountability.[146] Ideally, participatory science projects like the Drift Catcher help to build political power in otherwise-marginalized communities, in addition to generating useful data. A solid base of power and support is indeed essential for activist groups trying to participate in the regulatory process. Consequently, pesticide drift activists spend considerable time building that base—listening, organizing, training, campaigning, and building alliances with other social movements. I elaborate on those efforts in the next chapter.

Finally, pesticide drift activism highlights the lack of *capabilities* facing many agricultural communities that directly exacerbate the consequences of pesticide drift, such as the lack of pesticide drift incident response protocol, limited Spanish-language staff in public institutions, and inadequate pesticide exposure knowledge among health clinic staff members. Pesticide drift activists therefore have consistently advocated improving these institutions. Certainly, some important capabilities that many residents do not possess (e.g., legal citizenship and safe neighborhoods) extend well beyond the realm of environmental regulation, and given this, many pesticide drift activists are also politically engaged in addressing such problems (e.g., through immigration reform advocacy and participating in local politics).

Pesticide drift activists thus operationalize a notion of justice that differs distinctly from those that increasingly predominate mainstream environmental politics today. Pesticide drift activism, as Schlosberg and other scholars have observed of the EJ movement in general, works with the perspective that justice means combating inequalities, the oppressive social relations that cause them, gaps in participatory parity, and inadequate capabilities. Accordingly, EJ requires a strong environmental regulatory state that accounts for inequality, oppression, lack of participatory parity, and inadequate capabilities, and that works with other state agencies and community groups to redress these forms of injustice. That perspective is quite different from the communitarian assumption that justice coheres at the local scale (since communitarians assume that inequalities, oppression, and gaps in participatory parity and capabilities disappear there) and the libertarian insistence that justice requires unfettered markets (since inequalities, oppression, participatory parity, and capabilities are simply not priority concerns in the libertarian framework). The EJ notion of justice suggests that effectively addressing negative environmental externalities will require a state-society relationship that elaborates deeply on egalitarianism and that is distinct from the libertarian and communitarian ideas of justice that increasingly shape mainstream environmental politics today.

Conclusions

Though the agricultural and chemical industries insist that they are overly regulated, the history of pesticide policy and regulation illustrates that the state has never effectively addressed the problem of pesticide drift. In this chapter, I have discussed the ways in which regulatory agencies' work to address pesticide drift is thwarted by numerous material factors, including particular policy mandates, regulatory tools, funding, and industry influence—all of which help get pesticides to market and hinder health-protective restrictions on their use. Cultural structures in the regulatory arena also legitimize this inadequate regulatory response, including the stories that are widely told about pesticide drift, predominant notions of justice, and deeply rooted and widespread ideologies such as the modern notion of progress, the risk paradigm, and the pesticide paradigm.

Pesticide drift activists directly confront these material and cultural factors that undergird, exacerbate, and obscure the pesticide drift problem. Through a wide range of tactics and practices, pesticide drift activists convey a vision of justice that differs markedly from those that govern the state's work. Notably, they argue that effective and socially just solutions to environmental inequalities like pesticide drift will require accounting for and combating material inequality, oppression, inadequate forms of meaningful participation, and a lack of basic capabilities. As I discussed in this chapter, pesticide drift activists often invoke the precautionary principle as a widely recognized framework for helping to operationalize their theory of EJ. In chapter 6, I elaborate on what it might mean to put the pesticide drift activists' theory of justice—and the precautionary principle—into practice. First, however, we must address an unanswered question. Why is pesticide drift activism such a fledgling and relatively small group of activists? Given the long-standing, considerable, and growing interest in food and agriculture in the state, why is pesticide *drift* activism such a nascent phenomenon? Answering this question requires that we look at the history of alternative agrifood activism in California—a task that I take up in the next chapter.

5

The Alternative Agrifood Movement

The very definition of a sustainable system just can't include routinely poisoning workers and residents.
—David Chatfield, CPR

Industry and the state are not, of course, the only actors who play a role in addressing agrifood system problems like pesticide drift. Over the past fifty years, countless individuals and organizations have endeavored to change the agrifood system—to bring critical attention to the failures of both industry and the state to adequately confront the extraordinary array of social inequalities along with the environmental problems that stem from the practices, technologies, and networks through which food is produced, processed, distributed, and sold. These activists also work to constructively reform many of these problematic practices, technologies, and networks, and promote alternative agrifood systems that ameliorate many of those concerns. In this chapter, I describe these efforts and examine how well they have addressed the pesticide drift problem. Although this discussion focuses on California, the priorities of agrifood activists there are consistent with those of agrifood activists elsewhere throughout the United States.

I use the term alternative agrifood movement to characterize the collection of these activist attempts to critique and reform existing agrifood networks as well as build alternative ones. The alternative agrifood movement is an umbrella term that I use to represent the overlapping work of four realms of activism: the farmworker justice movement, antitoxics critics of conventional agriculture, alternative farming organizations (including sustainable farming and small farm advocates), and food reformers (a diverse category that includes nutritionist "what to eat" diet reformers, gourmet "foodies," and antihunger, food justice activists). The latter two—alternative farming organizations and food reformers—are the most active parts of the alternative agrifood movement today. This has not always been the case, however. In this chapter, I tell a history of agrifood activism to describe and explain the temporal shifts in agrifood activist priorities along with their associated consequences. I pay particular attention to the ways in which some of the primary forms of alternative agrifood advocacy have unwittingly abandoned and exacerbated the pesticide drift problem. This historical background and analysis will explain the emergence of pesticide drift activism and how its strategies and tactics overlap only partially with those of the rest of the alternative agrifood movement.

I should say a few words here about the terms I am using in this chapter. First, like all typologies, this categorization of the four realms of the alternative agrifood movement obscures the fact that they overlap. Many organizations and individuals, for example, promote alternative farming systems and press for food reform. Most activists do not define themselves so firmly into one of the categories that I have presented and are probably not accustomed to drawing such sharp lines between the realms, as I do in this chapter. That said, I delineate these realms, and highlight the tensions between them, in order to identify the strengths and limitations of the competing models of change and visions of justice that exist within the broader alternative agrifood movement.

Second, the reader should also recognize that my discussion glosses over the tensions that exist within each realm, since my goal is to critically evaluate some of the major trends within each. Some of these issues compel Patricia Allen (2004) to use the plural term alternative agrifood movements in her book *Together at the Table*. I use the singular term alternative agrifood movement for several reasons. Though divisions exist, the hundreds of groups that comprise the alternative agrifood movement share a key conviction that food system reform is important because food is qualitatively different from other commodities: it is necessary to sustain human life, its production and distribution have tremendous impacts on the planet's ecosystems, and it constitutes the primary livelihood strategy for a considerable portion of the global population. In referring to this movement in the singular, I do not mean to suggest a perfect cohesion and unity among alternative agrifood activists. Instead, my goal is to show that despite the fact that activists have drawn tremendous and much-needed attention to many problems in the agrifood system, several major injustices—notably, food justice, farmworker justice, and pesticide drift—remain relatively marginalized within the alternative agrifood movement. My discussion below shows how and why activist attention shifted in ways that marginalized these justice problems, and portrays what some relatively new groups of activists are doing to change those oversights. I pay particular attention to pesticide drift activists, examining how their work directly challenges the conception of justice governing many of the most popular forms of alternative agrifood activism.

The History of Pesticide Activism

Most histories of environmental activism in the United States emphasize early activists' preoccupation with wilderness preservation, and how debates about the environment typically focused on the tensions between preserving wilderness and conserving natural resources for perpetual harvest.[1] That said, progressive reformers of the early twentieth century addressed a number of urban and industrial environmental problems, including contaminated water, air pollution, garbage, and various workplace hazards.[2] During this time, pesticides and other chemicals also gained the public's attention at specific moments when found to be contaminating the food supply. Yet the public's worries about agricultural pesticides did not garner a significant public spotlight until the 1960s. Although agricultural community residents occasionally reported concerns about pesticide drift, the earliest sustained activism against the forms of pesticide exposure experienced by people living and working in agricultural communities emerged in the 1960s with a pesticide campaign led by the UFW.[3] My narrative about pesticide activism therefore starts at that point. Early pesticide activism suffered from numerous structural constraints but signaled a major change in public awareness of pesticide pollution in agricultural areas, as the section below illustrates.

Early Pesticide Activism (1960s–Early 1980s)

Some of the earliest and most lasting political activism against agricultural pesticide exposures took place through farm labor organizing efforts. Laura Pulido and Linda Nash have both shown that the UFW struggles of the 1960s and 1970s in California played a crucial part in the development of early worker protections from agricultural pesticide exposure, and Cole and Foster argue that the UFW-led pesticide campaign was "perhaps the first nationally known effort by people of color to address an environmental issue."[4] A key institution in the civil rights movement, the UFW pursued pesticide protections as an essential component of its overall effort to address the exploitation and vulnerability of California farmworkers through union contracts. With little pesticide regulatory apparatus to rely on and few possibilities for traction in the legislative arena at that time, the union adamantly argued that UFW contracts were the critical tool necessary for securing pesticide protections. To secure those contracts, the UFW engaged in a series of well-publicized lawsuits and broad-scale consumer boycotts. The boycotts were successful largely because they demonstrated the overlapping health problems that some widely used pesticides posed to both consumers and workers. Pulido insists that demands for pesticide protections were a useful tool for the UFW's overall project to "remedy the great power imbalance between growers and field workers," as they "served to attract the support of those who otherwise may not have supported the UFW."[5] The UFW's campaign materials from that period explicitly try to expand the union's base of support by connecting consumers and workers in terms of their shared exposures to pesticides (see figure 5.1 below).

Ultimately, the UFW's organizing, boycotts, and lawsuits led to a court ruling that pesticide reports are public documents, helped lead to a national ban on DDT, aided in illustrating the inextricable relationships between environmental problems and social inequalities, and secured union contracts (containing relatively elaborate pesticide protection clauses) for tens of thousands of California's farmworkers.

A number of other agrifood organizations filled key roles in this early period of pesticide activism and farmworker justice. Notably, the California Agrarian Action Project (CAAP) and other activist groups filed a lawsuit against the University of California in 1979, arguing that the university was using taxpayer dollars to fund research (on tomato harvest machinery) that would benefit large-scale farms to the detriment of small farms and farmworkers. Campbell describes how what was commonly called the "tomato harvester lawsuit"

came to be seen as a symbol of the broad choices facing California agriculture. Would public policy and university research continue to support conventional agricultural systems with their primary emphasis on economic productivity, or would they reflect a broader range of values, including environmental integrity, social justice, and the viability of family farms and rural communities?[6]

CAAP members also staged sit-ins in support of farmworkers put out of work by the tomato harvester.[7] The lawsuit and political activism eventually compelled the University of California to fund a Small Farm Center plus hire several Spanish-speaking extension agents. In the early 1980s, CAAP also played a major role in fighting for some of California's early key pesticide regulations, including the Birth Defect Prevention Act (SB 950, which requires the mandatory testing of pesticides for their potential to cause birth defects and other chronic health effects), the Right to Know

Figure 5.1
UFW poster from 1969 (and see Pulido 1996b, 106). *Source:* Courtesy of The Bancroft Library, University of California, Berkeley (BANC MSS 86/157c, Carton 21, Folder 21:9).

Act (AB 2033, which allows the public to review pesticide safety studies), and the Pesticide Contamination Prevention Act (AB 2021, which restricts pesticides that contaminate groundwater).

Yet most agrifood organizations' confrontational activism on farmworker rights and pesticide regulations was short-lived. To some extent this was due to the precipitous decline in UFW representation in the 1980s, as the union was plagued by its own internal problems, competition from other labor unions, the shift of political power in California's leadership to the Republican Party, growers' increased reliance on farm labor contractors, the growing power of agribusiness and its antiunion politics, and a tremendous influx of immigrants desperate to work for low wages and unable to participate in union activity.[8] This decline in power highlights one of the primary weaknesses of the UFW's pesticide campaign: its insistence on securing pesticide protections through individual union contracts rendered those protections temporary, vulnerable, and spatially uneven. Nonetheless, the UFW pesticide campaign of the 1960s (and its less coherent reemergence in the 1980s) left behind the important legacies of coalition building and conceptualizing economic inequalities as the root of environmental problems—a practice and perspective, respectively, that would come to influence pesticide drift activism after 2000.

The decline of the UFW and the institutional context that made organizing difficult for the union (such as the decline of the broader civil rights movement, and the rise of neoliberal regulatory reform and ideology) played an important role in undermining the efforts of agrifood activist groups to gain political traction on farmworker protections and pesticide reforms. At the same time, farmworker vulnerabilities increased but simultaneously became less visible. While people's needs to migrate have not declined, farmworkers have confronted increasingly hard-line U.S.-Mexico immigration and border policies along with hostile anti-immigrant sentiment. Immigrant farmworkers who are unauthorized—or those with unauthorized family members and friends—consequently must keep a lower public profile today than ever before. All of these factors contributed to the fact that agrifood activism started to take on a remarkably different form in the 1980s.

The Alternative Agrifood Movement Today

During a period in which political claims about civil rights for working people of color were supported by a larger social movement, California organizations were able to include these claims for justice in their agendas. As claims for these rights were replaced by neo-liberal arguments about individual responsibility, AFIs [alternative food institutions] withdrew from direct opposition to powerful political and economic structures and framed their programs in terms of the rights of consumers to choose alternatives, rather than in their rights as citizens.[9]

In this changing context, many key actors in the alternative agrifood movement throughout the United States have, over time, come to drop their involvement in oppositional, justice-based, and agricultural workplace-oriented work. In this section, I describe the predominant, most vibrant forms of alternative agrifood activism today. Their diverse array of approaches to agrifood system reform, as will become clear, reflects activists' reactions to a shifting institutional context and the influence of multiple notions of justice. This review also raises serious questions about the extent to which different forms of activism are able to address the very social and environmental problems in conventional agriculture that the alternative agrifood movement finds so problematic.

The two most well-funded and active realms of today's alternative agrifood movement focus specifically on farmers and eaters: alternative farming advocates, and food reformers. First, "alternative farming advocacy" describes two related types of work: small farm advocacy, and the sustainable agriculture movement. Small farm advocacy organizations lobby for policy reforms that protect small-scale, "family" farms that are increasingly disadvantaged by the larger forces of farm concentration, "scale-neutral" agricultural policies, and the corporate consolidation among off-farm food system actors like input suppliers, food processors, and food retailers. Small farm advocates also work to establish and promote the marketing and distribution networks that support small-scale farmers, such as organizing direct marketing venues like farmers markets that boost farmers' share of the final price, and developing labeling schemes to make small producers' and processors' products stand out and fetch a price premium in the marketplace.

Sustainable agriculture organizations' priorities include developing and disseminating less toxic agricultural production practices, developing alternative marketing systems that reward sustainable farming practice (e.g., organic), and lobbying for policy reforms that support sustainable farmers' interests. In terms of production practices, a wide array of organizations have long worked to develop and disseminate alternative farming practices that strive to reintegrate principals of ecology into agricultural production, develop less toxic pest management, reduce farmers' costs of production, and diversify farming practices to protect against both ecological and economic problems.

In practice, advocacy for small farms and sustainable agriculture over-lap considerably. Sustainable agriculture organizations often highlight their work with small-scale farms (often called "family farms"), small farm advocates frequently encourage less toxic agricultural production methods, and the alternative distribution and marketing systems that they nurture and advocate generally embrace both ecologically sustainable and small-scale farming. Notable California projects include the Community Alliance with Family Farmers' (CAFF) Lighthouse farm network, a farmer-to-farmer education program where farmers share information about economically viable and less toxic farming systems; CAFF's Biologically Integrated Farming Systems program, which provides technical assistance to help farmers reduce their use of pesticides; AMO Organic's farmworker-to-farmer training and small farm incubator program; and the Organic Farming Research Foundation, which supplies grant funding for organic agricultural research, and also works to "cultivate state and federal policies that help to assure the economic viability of organic family farmers."[10] Today, hundreds of such organizations around the world pursue similar projects, often working in collaboration with university researchers, extension agents, and at times, regulatory agency staff.

The organic food and farming movement is, without comparison, the most well-known and fully developed alternative farming production and distribution network, and its focus on pesticides makes it the one most directly relevant to the pesticide drift problem. The organic movement—in many ways the flagship of California's contemporary agrifood activism— is the most prominent and fastest-growing form of antipesticide politics.[11] Historians have shown that organic farming has many different social movement roots: health and natural food advocacy, the 1960s' back-to-the-land movement, and environmental activism.[12] Tim Vos emphasizes that organic "pioneers" pursued alternative socioenvironmental relations, demonstrating the viability of alternative production systems, protecting small-scale farmers, and "challenging the hegemony of agro-industrial paradigm."[13] Organic was solidified, standardized, and spatially expanded in North America by the USDA's development of the National Organic Standards, a codified list of allowable inputs, practices, and certification requirements that went into effect in 2002. Due in part to the ways that these standards were designed, but generally to the fundamental dynamics of economic competition, increasing land values, and already-existing corporate consolidation, the organic industry has morphed into a marketing niche dominated by growers seeking the label's price premium, is no longer tied to the small-scale farming of its movement roots, and is increasingly controlled by an ever-shrinking number of corporate processors and retailers.[14] Notwithstanding these changes in the organic movement, the growth of the organic sector has brought indisputable and critical attention to agricultural pesticide use. The public's concerns about pesticide residues on food have at specific moments constituted a powerful force in the legislative arena—indeed, they have driven meaningful interrogations of pesticide practices and health-protective pesticide regulatory decisions.[15] Consumers' increasing demand for organic food rewards environmental stewardship, has enabled more acres to be devoted to organic production, demonstrates the practicability and economic viability of less toxic pest management, incentivizes further research and development of organic production methods, and provides an avenue through which many consumers start thinking critically about food system ecology and politics.[16]

Second, food reform constitutes the other predominant realm of alternative agrifood advocacy today. As I write this book in 2010, public discussions of food reform are ubiquitous, and shaped overwhelmingly by a number of wildly popular "diet reform" books by the likes of Michael Pollan (*The Omnivore's Dilemma* and *In Defense of Food*), Marion Nestle (*What to Eat*), and Barbara Kingsolver (*Animal, Vegetable, Miracle*), the mainstream films they inspire or help to produce (*King Corn, Supersize Me,* and *Food Inc.*), and "locavore" organizations such as Slow Food. This particular collection of food reformers champion alternative shopping and eating practices, instructing the reader that growing one's own food, shopping locally, and eating "food, mostly plants, and not too much" (and especially not high fructose corn syrup) constitute *the* recipe for achieving numerous social goals: providing marketing opportunities for small farmers, boosting community economic development, rekindling a sense of community, solving the "obesity epidemic," and becoming knowledgeable about the agrifood system.[17] The propensity of food reformers to put particular stock in "buying local" as a sure way to solve seemingly any agrifood problem is evident, for example, in one local food blogger's recent claim that buying locally grown produce even constitutes an effective alternative to pesticide regulations:

> As the California Department of Pesticide Regulation makes their decision about whether or not to allow the use of methyl iodide, the rest of the country should stick to local food. Local food from small farms, organic or otherwise, is generally a safer, more sustainable bet.[18]

It is worth noting that other activists and scholars view this approach to food reform more skeptically, observing that it typically problematically frames individuals' food choice as the most important contributor to poor health; conceptualizes citizenship almost entirely in terms of shopping tips (where to buy, and who to buy from); dishes out unreflexively elitist lifestyle recommendations (e.g., grow your own food, cook three meals per day from scratch, and sit down and eat those meals leisurely with friends and family); fails to address the problem of hunger or its structural roots in economic inequality; ignores the material and structural supports of various eating practices; pays no heed to the many other health problems (and their causes) that are at least as problematic from individual and public health perspectives; and blames fat people for an extraordinary array of social problems.[19]

Community food security and other innovations in antihunger or food justice advocacy represent a different approach to food reform.[20] Community food security activists battle food insecurity (defined as inadequate access to nutritious food). They do so by emphasizing that food insecurity must be understood as a function of structural inequalities, rather than a consequence of individuals' decisions or conditions. The community food security movement emerged in the 1990s as a critical response to both state and private antihunger efforts that treat hunger as a temporary condition affecting individuals that is best solved by making food available in times of so-called emergency or crisis, and to the neoliberal backlash against "the undeserving poor" that has further undermined state food provisioning programs since the 1980s. These activists insist that affordable, accessible, nutritious, and culturally appropriate food be treated as a state-guaranteed human right.

Community food security organizations spearhead and support a wide variety of creative projects that actively aim to boost access to quality, affordable food in poor, urban communities along with other "food deserts" whose residents only have access to corner stores stocked with alcohol, tobacco, sugar, salt, and fat. Academic observers elaborate on the ways that activists have established farmers markets, community gardens, and market gardens in underserved communities; worked with local and state agencies to provide public grants that subsidize grocery stores in such places; and reformed public transit systems to connect residents of underserved neighborhoods to nearby, high-quality food markets. The community food security movement also includes a wide array of efforts to reorient government assistance away from being a dumping ground for the nation's surplus agricultural commodities and toward serving the nutrition needs of low-income people. For example, community food security organizations helped to establish the federal Farmers Market Nutrition Program that enables participants in the Women, Infants, and Children federal food assistance program to use food stamps at farmers markets. They have also reformed school lunch programs in many low-income communities around the country, often through creative efforts such as establishing salad bars in school cafeterias and developing farm-to-school programs that link schools with local farms. Although such projects are highly localized, umbrella NGOs like the Community Food Security Coalition have helped to coordinate the work of disparate local food security organizations. The Community Food Security Coalition plays an important role in coordinating efforts to lobby for national food and agricultural policy reform.[21]

In addition to increasing access to food, community food security projects often serve to promote social interaction, educate the public about nutrition and ecology, promote sustainable agricultural practices, support local farmers, rehabilitate marginalized populations, and provide jobs that develop skills in food production, processing, and marketing.[22] In the Food from the 'Hood project based at Crenshaw High School in South Central Los Angeles, for example, students run a garden, donate some of the produce to needy community residents, sell the produce at a local farmers market, process and sell a line of salad dressings nationwide, and use the profits to keep the project running and provide scholarships to graduating seniors.[23] The Homeless Garden Project in Santa Cruz offers job training and rehabilitative horticultural therapy and generates income through its produce sales.

With the critical attention it draws to the structural underpinnings of food inequalities, community food security represents a significant departure from the alternative farming movement and diet reformers. Although making major gains in terms of food consumption and the spaces of alternative food production, though, these predominant realms of alternative agrifood activism do not directly attend to the primary structural supports for the pesticide drift problem: the hegemonic pesticide paradigm, inadequate pesticide regulations, and the social inequalities and oppression experienced by immigrant farmworking communities.

Rather than pursuing pesticide regulatory reform that would curb pesticide use in conventional agriculture, alternative farming advocates and food reformers focus on developing markets (e.g., organic or local) that support alternative farmers, and advocating for policies that directly support those markets, alternative farming practices, and access to nutritious food. Based on interviews with thirty-seven leaders of prominent agrifood initiative organizations in California, Patricia Allen and her colleagues (2003) report that market-based and entrepreneurial programs dominated nearly all of those institutions' work.[24] They identify a range of material and cultural factors that drove this shift in activist practice away from environmental and labor regulatory reform. In addition to the previously noted structural factors that have made regulatory and policy reform increasingly difficult in the past thirty years, activist leaders reported that the shrinking pool of foundation funding on which so many advocacy organizations depend had become increasingly limited to entrepreneurial projects and often specifically excluded policy work. Activist leaders also reported that working on alternative projects is frequently more personally rewarding than earlier oppositional work.

Consequently, at the same time that the organic food industry, local food systems, and other alternative agrifood market institutions have expanded and matured, pesticide drift incidents and use rates of the most problematic pesticides have continued unabated. The forms of activism that are presently predominant in the alternative agrifood movement simply do little to address the inadequate environmental regulations that undergird pesticide drift. Empirically, the growth in acreage under organic and other alternative forms of agricultural production is offset by the continued use of highly toxic and drift-prone pesticides at extraordinary application rates. In other words, obscured in the shadow of organics' spectacular rise remain the recurring pesticide drift incidents as well as troubling data about pesticide use on the other 97 percent of California's agricultural land, such as the growing use of soil fumigants.[25] Although recent changes in pesticide activism have brought considerable attention to the health effects of pesticides, they have done so by squarely defining those effects in terms of consumer health and food safety. The alternative agrifood movement's focus on organic and other forms of alternative agriculture as a solution to the pesticide problem (and the simultaneous abandonment of regulatory reform) effectively privileged the bodies of relatively wealthy and well-informed consumers who purchase organic food, and abandoned the spaces of conventional agriculture and the bodies within them. As I will elaborate later in this chapter, pesticide drift activists—with their focus on cleaning up agriculture through pesticide regulatory reform—should thus be understood as filling the void left by the broader alternative agrifood movement's shift to developing and nurturing alternative food systems.

At the same time that California's agrifood activists shifted their focus toward food reform, entrepreneurialism, and the niche spaces of alternative farming, farmworker issues have largely been left behind. In her 2004 book, *Together at the Table*, Patricia Allen found that many major sustainable agriculture institutions at that time privileged the interests of alternative farmers and completely ignored farmworkers (and other non-farmer food industry actors).[26] This represented a shift from past practices, as many of the organizations that now concentrate almost exclusively on protecting small-scale farmers actually used to be centrally involved in farmworker justice advocacy. The history of CAAP (which evolved into the Community Alliance with Family Farmers, or CAFF) exemplifies this shift "from the stick of regulation" to "the carrot of incentives," and CAFF's current work centers exclusively on the needs of small farmers.[27] Many of the leaders interviewed by Allen and her colleagues reported feeling compelled over time to abandon their farmworker justice projects, because the farmers in the alternative marketing systems that they represented and worked to support actually depend on the same immigrant labor relations as conventional farmers.[28]

Additionally, the lack of attention to farmworker issues can be attributed to the agrarian ideal that is pervasive throughout agrifood politics and U.S. society in general: the widespread presumptions that small-scale, organic, or other alternative farms adhere to a family farm model; those farms do not rely on hired labor; and family farms represent the pinnacle of hard work, virtue, and the "foundation of democracy."[29] Wendell Berry's and Wes Jackson's writings exemplify this narrative. They are extremely popular among alternative agrifood activists and inspire popular agrifood writers like Pollan. Other scholars, on critically examining the pervasiveness of this idealization, have contended that it renders invisible the gendered division of labor, racial and ethnic discrimination, and violence that gave rise to the social and material relations from which agrarian populism emerged.[30] Critical landscape theorists like Raymond Williams and Don Mitchell have long shown how popular conceptions of rural landscapes as laborless, peaceful spaces depoliticize agricultural labor relations and render workers invisible.[31] Christy Getz and her colleagues argue that this misperception exists about the organic agriculture industry because the majority of organic farms are small (although relatively large farms do produce the majority of organic and other alternative food), and because small-scale farmers circulate widely in organic farming political circles.[32]

As a result, the image of small family farms gets considerable traction. The U.S. public tends to view farmers in general as fundamentally different from other employers—more moral or ethical—and thus deserving of exemptions from labor protections.[33] Getz and her colleagues write,

The idea that the family farm, in and of itself, represents a meaningful measure of social justice is widely accepted within alternative food movements. . . . Because growers are viewed as different from other employers, they are less likely to be held accountable for exploitative practices.[34]

Hence, when agrifood politics directly address social issues, the widespread idealization of the independent, family farmer means that justice gets conceived in terms of family farms' economic survival. The predominance of such ideologies helps to bolster food and farming reformers' support as well as sympathy for small-scale farmers, but not for the hired workers on which they depend.

The problem is that farmers' interests do not neatly align with those of their workers; in fact, in many ways the opposite is true. In their survey of organic farmers in California, Aimee Shreck and her colleagues found little support within California's organic farming community for the inclusion of social justice standards into organic certification criteria; only 25 percent of the organic growers they surveyed thought that organic certification should include *any* working condition standards.[35] Moreover, they found that organic growers are just as likely as their conventional counterparts to oppose basic worker protections, including minimum wage increases along with workplace health and safety regulations. In fact, 68 percent of the organic growers they surveyed stated that farmworkers should not have the right to collectively bargain—even though California law already guarantees it.[36] Getz and her colleagues also show that the organic farming community took a leadership role in fighting a California referendum that would have largely banned stoop labor, which is widely acknowledged to contribute to severe ergonomic pain and physical disablement.[37] Even though organic farmers would have been largely exempted from the legislation, they actively fought it in order to avoid relinquishing control of the labor process to the state. Although the public often assumes that organic farmers treat their workers better than conventional farmers do, that relationship is substantiated by neither empirical evidence nor logical argument.[38] In short, the most firmly established and presently popular approaches to alternative agrifood activism do little to confront the particular social inequalities as well as forms of oppression that operate as one key set of structural supports for pesticide drift.

The change that pesticide-specific agrifood politics have undergone since the 1960s is thus a profound one: from collective action that crossed class divides and interrogated injustices at the site of production, to individualized consumer politics whose environmental benefits are tied to shoppers' purchases.[39] Most important, pesticide politics have shifted from combating pesticide use in conventional agriculture toward targeting support for alternative, less toxic spaces of production. As a result, at the same time that labor relations have become the elephant in the room for the alternative agrifood movement, increasingly hostile immigration policy enforcement deepens immigrant farmworker vulnerability and invisibility; neoliberal reforms weaken pesticide regulations; and pesticide reform has become dislocated from the bodies of farmworkers, questions of farmworker justice, and the spaces of conventional agriculture itself. All of these problems are simultaneously exacerbated and obscured by the common assumption that alternative farming systems are more socially just than their conventional counterparts.

Although farmworker justice is not presently a priority for food reformers or alternative farming organizations, several new forms of farmworker justice activism have sprouted in recent years that deserve consideration. Their diversity of approaches raises interesting questions about the strengths and weaknesses of different methods to pursing social justice. For example, in recent years, several social justice certification schemes (such as the nascent domestic fair trade projects) have cropped up, bringing together activists from the various realms of the alternative agrifood movement.[40] Much like the organic and international fair trade systems, farmers participating in domestic fair trade certification schemes comply with various worker protections (and often, with some environmental protections), label their products accordingly, and receive a price premium in the marketplace. Many such schemes are limited to small-scale farmers. Domestic fair trade brings farmworkers into the fold of currently popular label-based solutions to agrifood system problems, and its rapidly increasing popularity stems undoubtedly from the fact that its mechanism of change is simple and accommodates consumer culture. Yet as is the problem with other label-based systems, domestic fair trade's social benefits are entirely contingent on consumers' interests and abilities, and thus generate benefits that are spatially and temporally uneven. Notwithstanding that limitation, the current propensity to advocate market-based solutions is so strong that many alternative agrifood activists propose this type of labeling scheme and shopping in (purportedly inherently just) local food systems as *the* appropriate solutions to seemingly any farmworker justice problem. For example, in response to a recent discovery that a group of Florida farmworkers was being enslaved, brutally beaten, and otherwise-egregiously abused and exploited by its employers, numerous advocates mobilized their concern by encouraging the public to buy fair trade and locally grown produce.[41]

For contrast, it is worth mentioning two different farmworker justice efforts that are designed to ensure more lasting worker protections. First, although farmworker union activity declined in the 1980s, farm labor unionizing is still active in the United States today. The UFW and the Farm Labor Organizing Committee (FLOC) have the largest memberships, and both unions continue to use consumer boycotts and large public demonstrations to help leverage farmworker union contracts. FLOC's approach illustrates interesting innovations in union organizing practice, in large part stemming from a recognition of the increasing concentration in the food-processing sector that makes growers subservient to processor demands. In addition to wage increases, grievance resolution protocols, health insurance, and other protections, some of FLOC's union contracts include provisions that processors cannot abandon growers in order to punish them for complying with worker demands. Moreover, FLOC contracts include foreign guest workers, who had elsewhere been used by growers to undermine unionizing activity and shunned by other unions.

Second, the Coalition of Immokalee Workers (CIW) is a Florida-based "community-based worker organization" (i.e., not a labor union). Emphasizing low wages that have not increased for thirty years, the CIW organizes consumer boycotts of fast-food restaurant chains and large-scale public demonstrations to leverage wage increases along with other labor protections for all tomato harvest workers whose employers supply those end retailers. Like the UFW and FLOC, the CIW builds creative alliances to develop a strong base of support (notably, with organizations such as the Student-Farmworker Alliance), organizes publicly through highly visible marches and other demonstrations (including, most recently, though its mobile Modern-Day Slavery Museum), creates its own short videos in which CIW members tell and share their stories, and enlists the support of celebrities (including members of Rage against the Machine). As of 2009, the CIW had successfully negotiated a penny-per-pound raise (which resulted in a 75 percent salary increase for workers) and other labor protections with some of the largest purchasers of Florida tomatoes: Yum! Foods (whose stores include Taco Bell, KFC, Pizza Hut, Long John Silver's, and A&W), McDonald's, Whole Foods, Subway, Burger King, and Bon Appetit Management Company (which runs food services for four hundred university and corporate cafés in twenty-nine states).

The CIW, the UFW, and FLOC enroll consumers in their activism, but they do so in a different way than domestic fair trade projects. Namely, like the UFW did in the 1960s and 1970s, the CIW and the worker unions use consumer boycotts as a way to enlist consumers to leverage *lasting* environmental and social protections for farmworkers. These protections are typically limited to a certain group of workers (i.e., those under union contract, or all harvest workers whose employers supply certain end retailers). Nevertheless, just as some of the UFW's protections were subsequently codified into California law, the CIW and the unions are currently pushing for regulatory protections for all farmworkers. For example, the UFW and FLOC actively press for comprehensive immigration reform, and the UFW regularly supports pesticide regulatory reform proposals and lawsuits. Also, in 2008, the U.S. Senate Health, Education, Labor and Pension Committee held a hearing on the the the CIW's penny-a-pound tomato campaign of 2008, and several senators later reported that they would introduce legislation to bring farmworker labor laws into line with those of the rest of the U.S. workforce.[42] What remains to be seen is whether FLOC, the CIW, the UFW, and other farmworker justice organizations will be able to successfully harness consumer power as well as engage in other forms of political activism to ensure lasting, universal protections for all farmworkers on a broader spatial scale (i.e., not just those in one labor union or supply chain, but throughout the entire United States or even internationally).

Theories of Justice in Alternative Agrifood Advocacy

In a trenchant and early critique, Patricia Allen and her colleague Carolyn Sachs argued for the need to bring social justice into the alternative agrifood movement.[43] Many observers, myself included, have repeated this call since that time. Yet I want to show in this book that it is worth thinking about justice in a more nuanced way. Rather than being absent, particular ideas about the meaning of justice have actively shaped these trends in alternative agrifood advocacy all along. In this section, I note the ways that some increasingly influential notions of justice have shaped the alternative agrifood movement in ways that exacerbate the pesticide drift problem.

Alternative agrifood advocacy, as I have shown, has shifted throughout the years as movement participants react to changing institutional and ideological contexts. In finding the path of least resistance, some of the predominant forms of alternative agrifood activism have come to rely extensively on market mechanisms and simultaneously abandoned the pursuit of regulatory reform. Organic agriculture, fair trade, and local food systems are all key examples of the ways in which some agrifood activists have embraced the market as the primary mechanism of social and environmental change. Though these activists might not be explicitly inspired or motivated by a libertarian conception of justice, they effectively invoke it through their mantras ("Vote with your dollars!"), favored projects (e.g., labeling schemes and direct marketing), and abandonment of environmental regulatory reform. Such work aligns with the prescriptions of libertarian politics: solving problems through increasing consumer choice in the marketplace, framing problems as individual rather than structural, and dismissing the need for state interventions into industry activity.[44]

This libertarian trend has been particularly evident in pesticide activism and poses problems for pesticide drift. The organic food system's pesticide protections are temporally uneven, since they are contingent on consumer whims, knowledge, and abilities, and are unconnected to any lasting or broader-scale environmental protections. When consumers have less ability or desire to pay for the organic premium, that mechanism for protecting people from pesticide exposure disappears. Additionally, alternative farming systems achieve pesticide protections that are inherently geographically uneven—limited strictly to those who work on or live near organic or other less-toxic farms—and are unable to deal with pesticide drift from nonorganic farms. Pesticide protections become a privilege of those who can afford to eat organic produce, or a lucky break for those few workers and residents who live and work near the organic farms that participate in the system. The price premium associated with the organic label also relies on the existence of barriers to entry for new organic farmers. As a result, the organic label provides an inherently limited opportunity for pesticide use reduction—if too much food is certified organic, supply outpaces demand, the price premium shrinks, and the incentive for farmers to participate disappears. The problem with the organic label is not in enrolling consumers per se but instead that the libertarian model of change is one in which social and environmental benefits are tied to isolated, individualized purchases, and thus are problematically undemocratic, contingent on ever-shifting consumer whims and abilities, enjoyed mostly by relatively privileged consumers, manifested unevenly across the landscape, and unbolstered by regulatory protections.

The last point is an important one. Those threads of the alternative agrifood movement that rely on market-based mechanisms invoke the libertarian notion of justice by abandoning the state as a crucial actor in protecting the public from problems like pesticide drift. Although many alternative agrifood initiatives evolved out of a frustration with an inadequate regulatory apparatus, the current focus on market-based solutions within pesticide-specific activism neglects the regulations that permit problematic pesticide use rates and application methods. In the flurry of market-based efforts, alternative agrifood advocates have largely ignored a crucial cause of the pesticide-intensive nature of conventional agriculture, and hence the most direct opportunity to change pesticide use patterns, pollution, exposure, and illness: pesticide regulations and regulatory practices. Like libertarianism more broadly, the alternative agrifood movement's reliance on market-based solutions to environmental problems effectively absolves the state of its responsibilities to ensure the conditions of social justice, and as such, unwittingly exacerbates the glaring problems (like pesticide drift) that were themselves deepened by libertarian-inspired regulatory reforms.[45] Moreover, relying on market-based mechanisms does little to meaningfully confront the social inequalities that constitute the root of regulatory failure: the extraordinary influence of industry over regulatory practices and decisions as well as the disproportionately low ability of agricultural community residents to do the same.

Despite the countless criticisms that can be made of it, the popularity of libertarianism and the market-based solutions it hails are so great that they almost seem unremarkable. Most of us now have a difficult time conceptualizing political engagement beyond our personal shopping decisions, in spite of the fact that such practices have a relatively limited ability to meaningfully solve environmental problems.[46] In that way, libertarianism functions ideologically, constituting a seemingly natural set of assumptions that reinforce the status quo.

In addition, considerable work in alternative agrifood activism explicitly and deliberately invokes a communitarian conception of justice. The communitarian argument is that communities should be free to determine what a good society is, and community members reach common understandings through tradition, shared experience, proximity, and relations of trust. The communitarian premise—that consensus, fairness, care, and representation cohere or at least increase at local decision-making levels— is particularly pervasive in food system localization advocacy. One of the universal assertions about local food systems, for example, is that direct relationships between producers and consumers (e.g., via farmers markets) cultivate relations of trust, transparency, and accountability. Although local food systems are generally not explicitly characterized as inherently just, they are framed as fostering justice, largely through enabling these personal connections between producers and consumers.

Food localization projects effectively create economic opportunities for small farmers, supply fresh local food for some consumers, and create spaces for public interaction. Yet in addition to the fact that no empirical evidence exists to link labor justice or environmental sustainability issues to farmers who market their products locally, researchers have found that some local political forums actually weaken and marginalize concerns about social justice and environmental externalities.[47] In contrast to the claim of community as inherently representative and inclusive, some local food movements and other localization efforts contain nativist elements and serve to further exclude marginalized residents and their concerns.[48] Local food advocacy is replete with a concern for fresh, healthy food, yet such advocacy often focuses on nutrition education while dismissing the structural causes of hunger and malnutrition.[49] Also, historically, environmental localization politics in California and elsewhere in the United States have historically served the economic interests of local elites rather than some generalized community interest.[50] Melanie DuPuis and Aaron Bobrow-Strain have shown that many food reformers have historically marshaled claims about "good food" to promote industrial growth, disparage "otherness," and craft "perfect," ideal citizens according to the norms of a white, privileged lifestyle.[51]

It is for such reasons that some agrifood scholars contend that the community built through farmers markets, the Slow Food movement, and other alternative food institutions frequently serves to reproduce, rather than confront, white privilege and other forms of inequality.[52] Although the notion has obvious appeal, the communitarian assumption that shared understandings about social issues exist at the local level is deeply flawed.[53] The point is not that conservative or industry-protective values necessarily dominate at the local level but instead that local politics cannot be *presumed* to foster progressive viewpoints or otherwise facilitate EJ. In short, the relationship between community and justice cannot be assumed; it must be explicitly interrogated and actively deliberated.

As Harvey notes, both communitarian and libertarian theories have a sufficiently "democratic and populist edge" to make them "potent political forces[s]."[54] The two often work in tandem as well. For example, many local food advocates claim that building local relationships naturally provides a pathway to social change that bypasses or replaces the state. This argument is exemplified in Pollan's recent book *In Defense of Food*, where he maintains,

Shake the hand that feeds you. As soon as you do, accountability becomes once again a matter of relationships instead of regulation or labeling or legal liability. . . . Regulation is an imperfect substitute for the accountability, and trust, built into a market in which food producers meet the gaze of eaters and vice versa.[55]

Though well intentioned, such calls situate justice for workers and the ability to avoid pesticide exposure as contingent on employers' whims along with consumers' desires to pay a premium in the marketplace—rather than as a set of guaranteed rights. The suggestion that proximity can provide a level of care that replaces the need for state intervention illustrates how communitarian and libertarian notions of justice can reinforce each other (and in quite problematic ways).

Appeals to the notion of community cannot confront problems like pesticide drift that have little to do with farmers' good (or bad) intentions, or relative concern for their surroundings and neighbors, especially given pervasive structural supports for the pesticide paradigm, the social inequalities and oppression facing farmworker communities, and scientific practices that systematically underestimate risk. Local food advocates' good intentions and commendable accomplishments do not weaken all the structural causes of social and environmental exploitation in agriculture. When people who suspect that they have been exposed to pesticide drift do not speak up for fear of losing their job or being deported, do not have the resources to substantiate or pursue their claims, or feel alienated from purportedly community events and spaces, their perspectives and experiences are sidelined as well as rendered invisible in local politics. In such contexts, the assumption that the local scale somehow engenders representation, transparent social relations, or environmental stewardship becomes dubious at best. Assumptions about community in fact obscure these significant disparities; Young argues that the concept of community "suppresses differences among subjects and groups."[56] To contribute to justice, food system localization will need to compel people to act as citizens beyond their purchasing decisions and to fight the regulatory, ideological, industrial, and other structural roots of social and environmental exploitation.

I want to emphasize that not all elements of the alternative agrifood movement are subject to these critiques. Indeed, some threads of alternative agrifood activism stray wildly from the communitarian and libertarian visions of justice. Community food security activism and other food justice efforts, for instance, constitute a notable case in point. Working from the perspective that accessible, affordable, nutritious, and culturally appropriate food is an essential capability for any community to function properly, community food security activists advocate for national food policy reform to boost the access to such food for underserved neighborhoods and social groups. They also work for a radical bolstering of national antipoverty programs in order to combat the structural inequalities that lead to food insecurity. The farmworker justice campaigns of the UFW, FLOC, and the CIW all explicitly prioritize redressing economic inequalities through securing lasting protections via agreements with buyers and employers, and fighting for reform of national and state labor and immigration laws. At the same time, these farmworker justice organizations' highly visible marches and other collective action events actively work to confront not only economic inequality but culturally oppressive social relations as well.

These critical threads of activism serve an important function in the broader alternative agrifood movement: they compel the public to understand food access and labor conditions as products of race- and class-based institutionalized inequalities, and they both illustrate and emphasize the need to combat them through collective action and confronting the state.[57] To the extent that such projects redress economic inequalities, combat oppressive social relations, and boost participatory parity and basic capabilities in disadvantaged communities, they align with the EJ vision of justice and thus contribute, albeit indirectly, to combating environmental inequalities like pesticide drift.

In contrast, the communitarian elements of agrifood advocacy argue that devolving political decisions to the community level is an adequate basis for justice, and the libertarian elements contend that the market is a sufficient mechanism for justice. Communitarian and libertarian perspectives thus ignore, reinforce, and obscure environmental inequalities. Sustainability practices and claims that draw on libertarian and communitarian theories are unable to effectively deal with environmental injustices like pesticide drift—problems that are rooted in inequalities, oppression, a lack of participatory parity, and inadequate basic capabilities—as well as their structural supports. It is for these reasons that pesticide drift activists pursue pesticide change through other means.

Emergence of Pesticide Drift Activism: Multiple Frames and Alliances

At the same time that alternative agrifood advocates have gained momentum and made notable strides in organic food, local food, and other alternative marketing systems, pesticide drift incidents continued unabated. Pesticide drift events—and the subsequent struggles to gain treatment, recognition, compensation, and future protections—made it clear to many residents and environmental activists that the pesticide drift problem needed to be confronted directly and explicitly.

Pesticide drift activism thus can be understood as having arisen in the wake of the various changes in other realms of the alternative agrifood movement. In this section, I discuss the ways in which pesticide drift activism emerges from, collaborates with, and also reacts to many different related forms of activism: the farmworker justice movement, the sustainable agriculture movement, the EJ movement, and air pollution activism. This examination will show that social movements with overlapping interests build strategic alliances but are never fully compatible. I look at how pesticide drift activists deal with such tensions, how they draw strength from alliances in spite of the contradictions, and how they manage coexisting yet competing problem frames.

Farm Labor Movement

Although many factors have weakened the agricultural labor union movement, it still plays a critical role in pesticide activism in California. Many pesticide drift activists had prior work with labor unions (including, notably, David Chatfield, executive director of CPR, and Gustavo Aguirre, a Central Valley community organizer for the Center on Race, Poverty, and the Environment). Additionally, the UFW and Lideres Campesinas (a farmworking women's organization in California) are both core members of CPR's steering committee.[58] Many pesticide drift activists are farmworkers, former farmworkers, or close relatives of farmworkers, or otherwise live in predominantly farmworking communities.

Still, without the possibility of a widespread farm labor unionization campaign, pesticide drift activists today focus less on workplace pesticide violations or protections, and more on the general movement of pesticides into nontarget fields, public spaces, and residential areas. Thus, their focus is more expansive than the 1960s' pesticide campaigns. Reaching out to residents engages a broader array of pesticide exposures as well as a more politically enfranchised constituency, and it enables activists to problematize the pesticide exposures exacerbated by farmworker injustice without being limited to the political quagmire of the farm labor market itself. Pesticide drift activists' concern for exposures beyond the workplace also illustrates that pesticide exposure cannot be controlled through personal protective equipment, reentry intervals, and other workplace-specific controls that regulators have tended to rely on.

Furthermore, because of the development of state and federal environmental regulatory agencies since the 1970s, pesticide drift activists presently see greater possibilities for broader regulatory reform than were imaginable during the UFW work of the 1960s. It should also be noted that pesticide drift activists focus on pursuing state- and federal-level regulatory reforms, since hard-fought battles at the local level generate only patchy protections. Pesticide drift activists therefore call on the regulatory state to ensure rights-based protections from pesticide exposure for all

California residents—not just those able to afford organic food, work on organic farms, effectively lobby their local officials, or be represented by a protective union contract.

Sustainable Agriculture Movement

Pesticide drift activism's relationship with the sustainable agriculture movement is best understood as an ambivalent one. The sustainable agriculture movement, reliant on organic and other alternative marketing systems to address the pesticide problem, does not really confront pesticide drift but does constitute an ally for pesticide drift activists. Pesticide drift activists recognize that practical and profitable less toxic pest management systems are, after all, part and parcel of the broader goal of reducing the use of the most toxic pesticides, and thus pesticide drift, exposure, and illness. Similarly, many sustainable agriculture advocates recognize that restricting the use of the most toxic pesticides is needed to compel many farmers to transition to less toxic alternatives. Around 2004, the major organizations working on pesticide drift decided that it was time to build more linkages with the sustainable agriculture movement. Having forged a broad and growing base of support in the Central Valley, and having won some notable victories in the regulatory and legislative arenas, they decided that augmenting their pesticide reduction efforts with work on sustainable agriculture would help frame and situate their critical, antipesticide efforts as constructive rather than antiagriculture. The alliance could also help make the issue of pesticide drift more visible to the sustainable agriculture organizations' broad consumer base, which consists of a large, growing, and politically enfranchised set of possible allies.

Pesticide drift activists have built several institutional linkages with the sustainable agriculture movement over the past few years. For example, CPR executive director Chatfield was on the board of the California Coalition for Food and Farming, one of the primary sustainable agriculture coalitions in the state.[59] In about 2008, CPR helped to bring various agrifood movement actors together in the California Political Action Network for Sustainable Agriculture—a nascent discussion forum in which major sustainable agriculture organizations come together regularly to share ideas for promoting sustainable agriculture via policy reform (including stronger organic food and farming policies, and enforcement thereof, as well as lobbying for Farm Bill reforms that increase subsidies for farmers who practice alternative pest management). CPR also structured its annual conference in 2010 to create a space for conversation among these activists about their overlapping interests; the conference theme was "Healthy

Harvest, From Field to Table: Bringing together farm workers, food lovers, pesticide experts, and community activists from across the state to learn, share, and break bread together."[60]

Pesticide drift activist leaders involved in these institutional alliances express three primary objectives in building bridges with these constituents: increase public knowledge about the ways in which agricultural pesticides move into the air, drinking water, and bodies of people living in agricultural communities; build a larger group of citizens who will hold the state accountable for enforcing and strengthening pesticide regulations; and make pesticide drift activism more constructive by linking it more explicitly with viable, less toxic alternative farming systems. Pesticide drift activists recognize that the sustainable agriculture movement's accomplishments help to make pesticide drift activists' regulatory demands economically feasible, just as increased regulatory restrictions on specific pesticides are crucial to making California agriculture more sustainable. Some activists find that sustainable agriculture advocacy is also a welcome change to confrontational, regulatory reform work. Describing the market-based projects and sustainable agriculture advocacy that some pesticide drift activists dabble in, one prominent pesticide drift activist told me in an interview,

[Using] the force of the market to drive change . . . would be so much easier than swimming against a raging current all the time. As a result, many of the people working on drift are actually working on this from the agrifood activist angle too. . . . I am so tired of swimming upstream. I got tired of fighting. It gets exhausting after awhile. I wanted to do something that would eventually take off on its own.

That said, what sets pesticide drift activism apart from the sustainable agriculture movement is that pesticide drift activists remain committed to pesticide regulatory reform and still spend most of their time fighting for the regulatory sticks that will force farmers to transition to less toxic forms of pest management. Regardless of their involvement in sustainable agriculture advocacy, pesticide drift activists exhibit an unwavering commitment to pesticide regulatory reform. My interview with a prominent PAN representative drove this point home. After an hour of conversation, including considerable discussion of the numerous "innovative" and "exciting" market-based projects that had recently piqued her interest, I asked her what she would like to see agrifood activists focus on during the next ten years. Her response was immediate and definitive: "Real, substantive FIFRA reform. And probably the key things that would be the goals would be precautionary language, [including] alternatives assessment."

It must also be recognized that fundamental divisions make the relationship between pesticide drift activists and sustainable agriculture advocates an inherently uneasy one. Pesticide drift leaders explicitly recognize that the sustainable agriculture movement's approach to the pesticide problem has largely benefited relatively affluent consumers principally concerned about the impacts of pesticides on their own bodies and those of their children. It was for this reason that one pesticide drift activist, during an interview with me, expressed ambivalence about the alliances she herself has helped to build with sustainable agriculture organizations: "There is a lot of potential, but we need to be smarter about how race and class and privilege work in order to build this linkage. There is a lot of race and class privilege that we need to confront."

Also, some of the prominent sustainable agriculture organizations do very little policy work, most focus their policy work on lobbying for stronger organic food and farming laws, and those few that are interested in politically engaging with pesticide drift policy have limited experience and skills for doing so. Organizations that advocate for small farmers refrain from participating in pesticide regulatory reform to avoid upsetting their nonorganic farmer constituents. Moreover, because the sustainable agriculture movement has prioritized the interests of farmers, who directly benefit from farmworkers' low wages and limited rights, many sustainable agriculture organizations refrain from advocating for worker protections.[61] Additionally, because the public tends to view organic as a product-purity standard (even though it is in fact simply a process standard), organic farmers fear that evidence of pesticide drift on to organic crops will threaten consumer demand for organic products. In other words, organic farmers—the core constituents of the sustainable agriculture movement—benefit from the invisibility of pesticide drift along with farmworker marginality, which are two of the structural factors that make pesticide drift such a pervasive problem in the first place.[62]

As a result of these tensions, the sustainable agriculture movement and pesticide drift activists now differ considerably in terms of tactics and style. As Allen and her colleagues found in their study of agrifood initiatives, earlier agrifood activists' "critical stance about conventional agriculture has more recently become subdued and framed as alternative rather than oppositional"—an orientation that does not suit pesticide drift activists.[63] In an interview with me in 2006, one early pesticide drift activist reflected that sustainable agriculture organizations' demonstrated preference for collaborative strategies such as outreach and farmer-to-farmer education projects did not account for the pesticide exposures that pesticide drift victims were experiencing.

The organic community—or the sustainable ag activists—are trying to expand the numbers of farmers who are practicing sustainable agriculture or organic production, and certainly the pesticide drift activists want that to happen as well. But how you get there is very different. . . . You'd think they would be natural allies but they weren't just because . . . the pesticide drift activists were so, I mean, *they were sick*! They were poisoned! And they were angry, and they were going to be confrontational. And the sustainable agriculture folks had a very different thing. They weren't *living* pesticide use like the pesticide drift activists were, so they weren't *compelled* in the same way to work on these issues as someone who's poisoned or whose child is poisoned. So strategies were really different, and messages were really different. And the sustainable agriculture community just wasn't as comfortable with what the pesticide drift folks were, whereas the pesticide drift people, I think they wanted the sustainable agriculture community to come on board, but it just wasn't going to happen.

This activist's next comments illustrate how divisive these differences can be:

They were pretty scared of us, I would say. Because we were calling farmers bad people because they were poisoning people. And that was not their . . . [Speaker pauses.] One, they didn't want to be political. Two, they didn't want to be associated with environmental wackos who were causing problems in the grower community. They had a different method for trying to get farmers to transition to sustainable agriculture. And certainly they had lots of successes, so an argument could be made that they were doing great work. But in terms of the politics of pesticide drift, they didn't want to be associated.

In interviews with me, other pesticide drift activists similarly express this conviction that cooperative tactics (with either industry or regulatory officials) have failed to protect residents from pesticide exposure, such failures indicate the need for confrontational tactics, and confrontation is an inherently constructive—rather than disruptive—component of pesticide reform:

People who are really at the grassroots level get so frustrated with politicians and with people, you know, who want to try the "sweetie pie" approach to the issues that affect our community on a daily basis. Because, you know, when you see people vomiting, and you see people semi-unconscious, and you see people out there going to hospitals after a pesticide exposure, you don't want to hear some shit about "We'll sit down in two months and talk about it." You know? No! That's not acceptable. (resident-activist)

Nobody challenged the farmers before, you know? In a big way. And now they're being challenged. . . . People kind of believe that you should follow the rules, be polite, and sometimes you can't afford to be polite. I've come to that conclusion. (resident-activist)

There's an issue about how confrontational you want to be . . . a lot of people who work for EJ groups, after the first time they heard, they were like, and you guys didn't do anything when they [regulators] wouldn't answer your questions? Are you crazy? That's an opportunity right there! *Roast them!* Because they're not going to change. And the only way they will change is if they're feeling pressure to, and if you're not applying pressure, then be prepared to have a lot more small, cozy-up meetings with them forever. (activist leader)

We're not a popular group! We're not very well liked! But that's because we're out there telling them the truth and making sure that the Central Valley is a better, cleaner place to live. (community organizer)

We are seen as pretty radical. We're fine with that. . . . People affected by pesticide drift have a much larger goal and vision. They're not willing to settle and compromise the way [less radical activist groups] are. And that's good. That's right. (community organizer)[64]

To be clear, pesticide activists themselves do not linger on such critiques, nor do they emphasize them publicly. Instead, they often conceptualize pesticide drift activism as the stick that is needed to push industry and research toward sustainable farming systems—complementing the organic industry's carrot of price premiums in the marketplace. Pesticide drift activists' efforts in recent years to forge alliances with the sustainable movement demonstrate their constructive recognition that meaningful reform in the agrifood system will require both market incentives and regulatory restrictions. Technically and economically viable alternative agrifood systems make pesticide regulatory and policy reforms politically palatable, and those regulatory and policy reforms are needed to compel farmers to adopt alternative technologies.

Yet pesticide drift activists' alliances with the sustainable agriculture movement are necessarily limited, contingent, and cautious. In an interview with me, one PAN representative reflected on the tensions that "will always be there" between the carrot-and-stick models of change:

We do not historically have a good relationship with growers because we have been all about the stick. . . . Until you get rid of the tools, there's no incentive for farmers to change. On the other hand, it is really hard for us to both be the stick, which we continue at the regulatory level, especially federal but also within California, [and the carrot]. . . . It's really hard for us to do both. But we want to do both, because we know full well and absolutely that the only real solution is sustainable agriculture. . . . So we have to play both, and it's a real challenge.

These tensions are far from disappearing. Another activist recently commented to me about the "rift between people primarily promoting the market-based approach and the confrontational/regulatory reform approach" among the participants in a new national working group of agrifood NGOs.

While bridging the sustainable agriculture movement and pesticide drift activism is both laudable and important, the shape that these linkages should take is still an open question. Will sustainable agriculture organizations really push their constituents to hold regulatory agencies accountable for pesticide drift and worker justice? Will sustainable agriculture organizations start to publicly recognize the limits of market-based and localist forms of change? Will they meaningfully confront the race and class privilege that shape their current practices and recommendations?

Some pesticide drift activists view these questions as fruitful opportunities for dialogue and critical reflection. For example, in reference to an article in the San Francisco *Bay Guardian* in 2009 that explicitly challenged the Slow Food movement's elitist tendencies, one pesticide drift activist leader told me, "Those are the types of discussions we should be having. Only in confronting those tensions will we find some resolution. We think conflict is good. Without it, the status quo will remain." Other pesticide drift activists are decidedly more cautious and frame their relationship with the sustainable agriculture movement in instrumental terms—reaching out to relatively wealthier constituents in order to build a larger group of citizens who will hold the state accountable for enforcing and strengthening pesticide regulations. Whether the sustainable agriculture movement can meaningfully help to address problems like pesticide drift depends on its ability to recognize that what is good for farmers is not always good for farmworkers (and vice versa), that risks to consumers do not map neatly onto the hazards endured by agricultural community residents (since many of the pesticides that are the most frequent contributors to pesticide drift do not pose major risks as residues on food), and that the public will need to do more than change its shopping habits (as pesticide drift will not disappear through markets and/or local marketing relationships alone). In any case, these tensions and uncertainties compel pesticide drift activists to refrain from placing all their eggs in the basket of sustainable agriculture activism. Instead, as I discuss below, they have formed stronger alliances with California's EJ organizations.

Environmental Justice

The EJ movement's focus on the relationships between social inequalities, oppression, and pollution resonates with many pesticide drift activists. In interviews with me, pesticide drift activists highlighted the ways in which various forms of social inequalities and oppressive social relations experienced particularly by farmworking families and others in agricultural regions—poverty, job insecurity, systemic dismissal and disregard by authorities, language barriers, and the lack of access to health care and legal assistance—prevent adequate regulatory response to, medical treatment of, and analysis of pesticide drift:

> To me it's very much not only exposure but what resources do people have after they're exposed? Who can go to a doctor? And who's worried about getting deported? And who's, you know, if they say anything, are fired or losing their jobs from exposure? So you can be disproportionately impacted by power dynamics too, by all these social structures in terms of health care and immigration and a bunch of different things too. (activist leader)

> Most of the communities, especially specifically in these accidents, are low-income communities of color, and communities that have very little political clout. A lot of them are migrant farmworking communities. . . . [Our concern is] the exploitation of these communities that haven't been able to protect themselves in the past. But we hope to empower them, that's our goal, is to educate and empower. (resident-activist)

In response to my question about how pesticide drift activism relates to the EJ movement, one CPR leader stated,

> That was actually another question that's been the subject of discussion at the beginning of the drift campaign and just in general within CPR overall. What does EJ mean, how do we work on it, how is it incorporated into our campaigns? . . . How much do you integrate versus compartmentalize and ghettoize and segregate [it]? . . . The answer is always, it should be in everything! There are seven hundred thousand farmworkers across the state who are predominantly people of color. They're adversely affected. They are unquestionably disproportionately affected. . . . At one point I even heard a comment, "Yeah, but aside from the farmworkers?" I [replied], "There is no 'aside from the farmworkers!'" . . . Overall a great number of farmworkers . . . fall into some "low-income people" and/ or "people of color" category, and that's a significant weight in that category, so there is no taking that aside.

Many pesticide drift activists push this EJ framing to the foreground in their political demonstrations (see figure 5.2).

Framing pesticide drift as EJ highlights the ways in which social vulnerabilities deepen the problem even as they render it invisible, and activists' work with residents of farmworking communities throws these relationships into particularly sharp relief. For pesticide drift activists, pesticides are only one injustice among many problems that disproportionately burden their communities. As a result, their pesticide drift work often is just one component of a long list of projects. The Grayson Neighborhood Council, for example, has fought not only pesticide drift but also

Figure 5.2
Activists in Fresno, California, 2007

a massive tire fire, waste incinerators, landfill expansions, nitrate- and arsenic-contaminated drinking water supplies, and a Superfund site in its small, predominantly Latino community. Its work is not limited to disproportionate environmental burdens but also includes its disproportionately low access to environmental goods; accordingly, the council successfully fought to get a community center, park, playground, and bicycle path built in its community. Reflecting on the range of issues she herself had been involved in, one resident-activist told me, "The bottom line is, it's kind of the same stuff, even though it might be different issues. It's people who seem to have an advantage who are taking an advantage. We feel like we need to stop that, because that's why we're suffering."

In accordance with this way of understanding the problems they experience, pesticide drift activists have many alliances with EJ groups and initiatives. Many of the core members of CPR's steering committee represent EJ organizations. Additionally, pesticide drift activists helped to develop the Central California Environmental Justice Network, whose conferences have functioned as the springboard for regional EJ projects and alliances. Pesticide Watch's new statewide director was trained in the Green Corps field school for community organizing and developed his skills through working with the Toxics Action Center in New England, helping community groups there fight hazardous waste. Pesticide drift activists continue to play a strong role in shaping the California EPA's EJ program and DPR's EJ pilot project in the Central Valley.[65]

In their observations of the broader EJ movement, Cole and Foster attribute its successes to strong community organizations along with regional alliances and coalitions—networks that "allow local groups to support each other, share strategies, and bring their combined power and resources to bear on local issues."[66] Pesticide drift activism similarly reflects this model of political activism, and pesticide drift activists emphasize the importance of building power within disadvantaged communities "from the ground up." Accordingly, CPR—the hub of pesticide drift activism in the state—was intentionally designed as a coalition rather than an organization. CPR's board and staff work to help channel resources toward community-based groups, rather than doing the work for them. In an interview with me in 2009, Pesticide Watch's current statewide director, Paul Towers, stated that this focus on grassroots, community-based work has always characterized pesticide drift activism in California, which was originally cohered through the work of six Pesticide Watch community organizers who sought "to support community groups on the ground, in the living room, in the community center." Towers described his involvement in pesticide drift activism as "place-based, community organizing," which can show that "we can change power structures" regardless of the issue at hand. He views community organizing as part and parcel of Pesticide Watch's two main goals: "to shift the entire framework of how pesticides are regulated in California and the U.S.," and "build leaders and advocates beyond the issues of their own backyards and take on whatever the challenge might be." The Center on Race, Poverty, and Environment, a core organization in pesticide drift activism, likewise expresses this commitment to community organizing in its Web site's description of its "three ambitions":

First, that individuals taking part in a particular campaign leave the campaign with more personal capacity than they had coming into it. Second, that the community involved has more power vis-à-vis decisionmakers at the end of the campaign than at the beginning. Finally, to concretely address the environmental hazard at hand.[67]

Chatfield, CPR's executive director, agrees that community organizing is "the best long-term solution to problems like pesticide drift"—a sentiment echoed by Tracey Brieger, CPR's campaign director:

How is there going to be some substantial change on drift? A huge part of the answer to me is a strong base of people who are interested, people who are going to be relentless, and who are going to be bold at fighting and fed up and really making the change. . . . The [premise of CPR] is: keep the resources out into the groups, keep the capacity going in the groups.

Those local groups in turn focus on community organizing. As one resident-activist in Lindsay stated in an interview with me, "We don't need to have money to have power. We need to have unity, especially in the small communities. Because with unity, that's all the power we need, and . . . we get unity through organizing."

Pesticide drift activists augment their grassroots capacity-building work by engaging in expert-driven legislative, regulatory, and litigation work, which is coordinated by a small number of central staff members of the major NGOs involved in pesticide drift activism. The following statement that a CPR leader made in an interview with me illustrates that pesticide drift activists actively struggle to keep these two tactics balanced:

I think we have to question ourselves for getting so comfortable with very few people within policy groups and CPR core staff having that much access [to regulatory officials] at the expense of other people having access. And for me, one of our main jobs is to open that up, is to use our access to make sure that other people do too. . . . With all due respect, I don't think it's [national NGO staff members] that they should be accountable to. As part of a larger group, yes, but not at the expense of people who are really being affected.

The regional focus on the Central Valley counties of Kern, Tulare, and Fresno by both CPR and the Center on Race, Poverty, and Environment was inspired not only by the area's disproportionately dangerous pollution levels but also by the fact that many nascent community groups there had explicitly requested technical assistance and guidance in organizing and pressing for political change. Staff members from the two organizations work to help local residents gain access to funding and decision makers, develop confidence in their public speaking skills and opinions, and otherwise generally become more politically powerful in their communities on a wide range of issues.

Though crucial and central to pesticide drift activism, grassroots capacity building is time-consuming, slow, and emotionally challenging. It is also rarely funded by the charitable foundations that activists rely on for material support; moreover, funders typically require organizations to demonstrate material accomplishments after one or two years as a condition for receiving additional funding—a time frame that simply does not suit long-term capacity-building work. While they struggle through these material constraints, activists continue to stress the significance of collaborations between groups within the coalition. As revealed by the following statement by one pesticide drift activist I interviewed, the coalition facilitates the process through which groups with political experience and resources help advise, motivate, and morally as well as materially support less experienced groups wanting to take on related environmental problems.

It is very frustrating work, because you are fighting systems. You are fighting entrenched systems and politics, and you know you are. It's a never-ending thing. It's one problem after another, and so at times, being human, you get discouraged. So when you see somebody that makes an effort to come a long way to back you up and to give you moral support . . . it gives you a little bit of juice to keep going. You know, that is really, really the importance of it . . . that it gives you that little bit of oomph to say that, "Well, yeah, I'm not going to let these assholes get me down, you know, I'm going to keep going, you know." And that is really the key part about the networking thing.

Air Pollution
Some pesticide drift activists question the value of framing pesticide drift as an EJ problem, arguing that emphasizing the racial and class dimensions sidelines the issue as a farmworker problem, and alienates the Central Valley's more politically enfranchised (and urban) residents concerned about air pollution but not agricultural pesticide use per se.[68] As a result, pesticide drift activists also frame pesticide drift as an air pollution

problem and pursue alliances with California's burgeoning air pollution movement, for which the influences of class and race do not play nearly so central a role in framing and fighting polluted air. Air pollution activists concentrate their fight on ozone and particulate matter, two of the basic "criteria pollutants" governed by the Clean Air Act and whose ambient concentrations in the Central Valley are among the highest in the nation and regularly exceed federal standards. One of the Central Valley air pollution activists' major concerns is with the connections between the area's well-substantiated air pollution problems and exceptionally high incidence of asthma (especially among children). Ozone, particulate matter, and other air pollutants stem not only from pesticides but also from cars and trucks, irrigation pumps, agricultural and residential waste burning, agricultural dust, dairies (hence the reference to the dairy industry's "Got Milk?" ad campaign in figures 5.3 and 5.4), and other sources.[69]

Participants in the fight against air pollution in the Central Valley include many grassroots organizations, but also relatively mainstream environmental organizations like the Sierra Club and the American Lung Association that shy away from radical frames like EJ—an eclectic mix that generates a variety of practices for engaging with air pollution regulators. The grassroots groups, which also participate considerably in pesticide drift activism and are more willing to highlight the EJ dimensions of environmental problems, accordingly engage in oppositional tactics that the mainstream environmental organization participants do not (such as rallies like the ones depicted in figures 5.3 and 5.4). As portrayed also in figure 5.4, air pollution activism constitutes an increasingly powerful multilingual and multicultural force. It is an approach with which pesticide drift activists build strategic linkages and broaden the ways that the public understands pesticide pollution.

Framing pesticide drift as air pollution broadens the scale at which pesticide drift is understood, thereby releasing the problem from the discursive constrictions of regulatory officials' accident framing, and highlighting the need for statewide or federal regulatory response. Many of the organizations and individuals involved in the Central Valley's air pollution movement also participate in pesticide drift activism (and vice versa). Through this alliance, and a mix of grassroots advocacy and lawsuits filed on behalf of community-based groups against regulatory agencies, the Center on Race, Poverty, and Environment and other activist groups successfully forced California regulatory agencies to end the agricultural industry's historical exemption from the Clean Air Act. A major component of this work was forcing regulatory agencies to recognize and reduce volatile

Figure 5.3
Activist rally with television news camera crew in 2007 in Fresno, California, outside the office building where state and regional environmental regulatory officials were meeting about air pollution.

Figure 5.4
Braulio Martinez, of Alpaugh, California, holds a sign during a rally outside a public hearing to call for stronger health-protective measures against pesticide-related smog at the Kearney Agricultural Research Center in Parlier, California. The sign reads, "We don't want any more contamination." *Source*: Photo taken on August 14, 2006 in Parlier by Teresa Douglass (*Visalia Times-Delta*).

organic compound emissions from pesticides, as these emissions produce ground-level ozone, a major contributor to asthma, other lung diseases, strokes, and heart attacks.[70]

While this alliance has been incredibly effective, activists involved in both pesticide drift and air pollution activism explicitly underscore that the two are, in the words of one CPR leader, "distinct [but] mutually supportive networks." The frame of air pollution, although able to reach a broad audience by presenting pollution as "everyone's problem," can gloss over the various forms of inequality and oppression that deepen and obscure the pesticide drift problem. Also, the air pollution movement tends to focus attention on criteria air pollutants (such as ozone and particulate matter) rather than specific pesticides. Scientific research on air pollution tends to sideline the contribution of pesticides as well; for example, scientific research on particulate matter, one of the most prevalent and highly regulated types of air pollution in California, does not distinguish various particulates by their chemical composition (only by their size), thereby hiding the extra toxicities posed by pesticides. Additionally, pesticides are generally regulated by different agencies than are other air pollutants, and agencies have used the lack of clear regulatory responsibility for pesticides (as air pollutants) to delay implementing new regulations (as occurred in California with the new regulations that require reducing volatile organic compound emissions from pesticides).

How the pesticide drift activists resolve the tensions between the various framings and alliances remains to be seen, as do the consequences of those tactical decisions. In the meantime, the advocates employ various framings, strategically using them to build coalitions to gain political traction and broaden the public's understanding of pesticide drift as an everyday problem requiring the state to implement precaution-based solutions.[71]

Theories of Justice in Pesticide Drift Activism

Reflecting on these various alliances provides a timely opportunity to consider the strengths and weaknesses of different activist tactics along with the underlying theories of justice that they align with. Pesticide drift activists' relationship with the sustainable agriculture movement is a notable case in point. As I discussed earlier in this chapter, pesticide drift activists understand and strategically develop alliances with the sustainable agriculture movement, yet they have been largely marginalized by its preoccupation with pursuing change through the development of organic, local, and other alternative farming methods and marketing systems, and its abandonment of the state as an agent of environmental change. This relationship between these two overlapping realms of alternative agrifood activism—and pesticide drift activists' emphasis on striving for change through the regulatory and policy arenas—in turn highlights some of the limitations of not only the sustainable agriculture movement's favored tactics but also the conceptions of justice that undergird them.

Sustainable agriculture advocates' and diet reformers' focus on libertarian-inspired, market-based solutions may largely be a strategic reaction to the broader political opportunities constrained and shaped by neoliberal regulatory reforms. Yet their *reliance* on a market-based model of change dangerously accommodates the neoliberal agenda, making protection from pesticides a privilege for those who can find and afford organic food, and effectively absolving the state of its responsibilities to ensure environmental protections for all people. At the same time, the benefits of market-based reforms accrue unevenly, providing benefits to some (in this case, relatively wealthy consumers and participating farmers) and abandoning others (in this case, farmworkers and residents of agricultural communities).

Similarly, to the extent that sustainable agriculture proponents and food reformers rely on food system localization and shortened supply chains to ensure environmental protections as well as socially just labor relations, they embody a communitarian notion of justice. In asserting that local communities constitute the locus of justice, care, and shared understandings, the communitarian ideal flattens local politics, dismissing and obscuring the extraordinary inequalities that persist at local scales along with the structures that reproduce them. To the degree that communitarian-inspired solutions endorse voluntary and devolved alternatives to broad-scale environmental and labor protections, they in turn also effectively accommodate neoliberal regulatory rollback and other failures of the regulatory state.

My point here is not that the sustainable agriculture movement does not do important work—it most certainly does, especially in terms of honing alternative pest management systems; demonstrating how and where they can be economically viable, technically practical, and ecologically beneficial; and justifying the call for increased research funding and federal subsidy money to be allocated to sustainable agricultural production. Likewise, as I discussed earlier in this chapter, food reformers pursue numerous critical objectives. Rather, what I want to emphasize is that the libertarian and communitarian logics that largely structure sustainable agriculture advocates' and some food reformers' current foci effectively absolve the state of its responsibilities, end up sidelining other problems (notably, pesticide drift), and unwittingly exacerbate the oppression experienced by farmworkers and other marginalized agricultural community residents. By assuming that the development of alternative farming and marketing systems can adequately address agricultural pesticide problems, well-intentioned advocates further marginalize and depoliticize the pesticide drift issue, its structural underpinnings, the spaces in which it unfolds, and the bodies within those spaces.

Moreover, when advocates' recommendations are limited to urging consumers to change their shopping and eating habits—as is often the case—they effectively dismiss the need for state intervention. Even while pesticide drift activists champion alternative labeling systems and other market-based solutions, they also recognize these limitations and are quick to specify how their own objectives differ. As one pesticide drift activist told me in an interview, "Local and organic are such a small part of the solution. . . . They are not the be all and end all of sustainable agriculture. . . . Is the best way to do that through individual purchasing? No. It's all about policy." Pesticide drift activists do not believe that localizing food systems can create better social or environmental relations in agriculture; in fact, the notion seems to strike most of them as oddly presumptuous. For example, after I asked one pesticide drift activist about the potential of food system localization to address the pesticide drift problem, she paused and then replied, "Even if the food is grown nearby, that does not mean it is grown without harming somebody." Pesticide drift activism shows that libertarian and communitarian ideologies do not help us deal with environmental inequalities like pesticide drift.[72]

Instead, pesticide drift activism works with a concept of justice that starts by fighting oppression, material and procedural inequalities, and inadequate basic capabilities. This orientation stems not only from pesticide drift activists' critical reaction to the state, as discussed in chapter 4, but also constitutes a reaction to the strengths and limitations of the other social movements with which they interact. In many ways, the labor justice and EJ movements have particularly influenced pesticide drift activists.

Pesticide drift activism specifically asserts a need for distributive justice within the broader alternative agrifood movement. Implicitly critiquing the sustainable agriculture movement and diet reformers for abandoning agricultural community residents in their preoccupation with farmer livelihoods and relatively privileged consumers' bodies, pesticide drift activists argue that socially just agrifood networks must protect *all* bodies and spaces—including those that bear a disproportionate burden of pesticide contamination. Pesticide drift activists' concern for the distribution of activist attention parallels the EJ movement's critique of the mainstream environmental movement for the latter's focus on wilderness, rather than the spaces where people live, work, and play, and for ignoring the disproportionate impacts of environmental harms on the poor and people of color.

In addition, pesticide drift activism illustrates that activists aiming to understand and effectively address environmental problems must recognize and protect against the processes through which various forms of oppression directly exacerbate environmental problems (and ultimately, must also work to ameliorate some of those forms of oppression). What brings pesticide drift activists together and compels them to build alliances with the EJ movement is indeed their identification with each other along one or more of the multiple, overlapping axes of oppression—poverty, racism, a lack of legal status, structural status in the agricultural workplace (i.e., for farmworkers), a lack of formal scientific expertise, and an ideological adherence to the precautionary principle rather than the pesticide paradigm—all of which politically marginalize them and strengthen pesticide-intensive agriculture. Pesticide drift activism also should be understood as an implicit reaction to the sustainable agriculture movement's history of ignoring these relations of oppression and, at times, actively fighting to maintain some of them.

Pesticide drift activists also contend that socially just activism must intentionally recruit and enable the participation of marginalized groups. In other words, justice requires bringing those who are most affected and least represented into the movement. Pesticide drift activists therefore tend to hold their meetings and conferences in agricultural communities, and at times when most agricultural community residents will be able to attend (i.e., on weekends and evenings), plus they typically provide translation, transportation, child care, and meals at those events. More broadly, this concern for participatory justice also shapes the way that pesticide drift activists design and manage the responsibilities of the core institutions. This is particularly evident in the fact that CPR's central purpose is to channel resources out to community groups, and help them gain access and influence vis-à-vis industry, the state, and activists alike.

Finally, pesticide drift activists maintain that justice requires social movement activity to help develop marginalized communities' basic capabilities, not just augment farmers' technological tools and reform consumers' shopping habits. Pursuing truly sustainable agrifood systems requires that all participants have the basic capabilities (such as time, knowledge, access to local officials, adequate transportation, background information, and other resources) needed to actively engage in social movement activities. It is worth noting that agricultural community residents and farmworkers who have become politically active in the fight against pesticide drift do not neatly compartmentalize pesticide illness separately from the other problems in their neighborhoods and regions that they want to confront, such as contaminated drinking water, a lack of textbooks in the local schools, irresponsible local officials, low voter turnout, and drug dealing, domestic violence, and other crime. Many individuals and organizations involved in the struggle against pesticide drift also work to build capabilities through many other activist institutions, such as farm labor unions, immigration reform advocacy, and voter registration campaigns.

In terms of its theoretical underpinnings, pesticide drift activism aligns closely with the EJ movement. Pesticide drift activism is highly influenced by the EJ movement's underlying theoretical framework and nurtures many important institutional connections with it. These ideological foundations that compelled the EJ movement to critique the mainstream environmental movement in the first place—the latter's blindness to issues of distribution, recognition, participation, and capabilities—similarly drive pesticide drift activists to reject some of the problematic tactics of other activists. That said, for a couple of reasons, it would be a mistake to conceptualize pesticide drift activism as situated perfectly within the EJ movement: some activists worry that the frame excludes some constituents and could brand the problem as marginal rather than pervasive, and pesticide drift activism encompasses many of the mainstream environmental movement's hallmark practices (such as expert-driven lobbying and litigation at the federal level). Notwithstanding these differences, pesticide drift activism illustrates the key components of justice articulated so well by the broader EJ movement: that justice requires combating distributive inequalities and cultural oppression as well as boosting participation and capabilities for marginalized groups.

Conclusion

In this chapter, I have described the ways in which some of the predominant realms of alternative agrifood activism have, in recent years, increasingly relied on local food systems, the organic label, and other alternative marketing systems to try to address pesticide problems in the agrifood system. In the process, they have largely neglected the pesticide drift problem and the people who live with it on a daily basis. These predominant forms of alternative agrifood activism increasingly align with the prescriptions of libertarian and communitarian notions of justice—that is, solving problems through market mechanisms and localization projects and abandoning the state. Driven by the broader neoliberal context and compelling rhetorical claims, these conceptions of justice do little to help activists meaningfully address pesticide drift and its structural supports.

Pesticide drift activism can be understood as a reaction to these trends in the broader alternative agrifood movement. To fight the problem of pesticide drift, pesticide drift activists nurture strategic but ambivalent alliances with the farmworker justice movement, the sustainable agriculture movement, EJ groups, and air pollution activists. Through these alliances along with their various tactics and practices, pesticide drift activism operationalizes a conception of justice that is distinct from those that presently govern the work of food reformers and sustainable agriculture advocates, and yet reflects in no small way pesticide drift activists' strong ties to the EJ movement—namely, a notion of justice that calls for fighting environmental inequalities, recognizing and dismantling oppressive social relations, improving participatory parity, and building basic capabilities in otherwise-marginalized communities.

6

Conclusion: Taking Justice Seriously

In the years ahead, I want to see a full-scale revitalization of what we do and how we think about environmental justice. This is not an issue we can afford to relegate to the margins. It has to be part of our thinking in every decision we make.
—EPA Administrator Lisa Jackson, 2009

The degree of risk to human health does not need to beat statistically significant levels to require political action. The degree of risk does have to be such that a reasonable person would avoid it. Consequently, the important political test is not the findings of epidemiologists on the probability of nonrandomness of an incidence of illness but the likelihood that a reasonable person, including members of the community of calculation, would take up residence with the community at risk and drink from and bathe in water from the Yellow Creek area or buy a house along Love Canal.
—Richard Couto, "Failing Health and New Prescriptions"

Pesticide drift is like so many environmental problems today: diffuse, elusive, hazardous, and invisible. As demonstrated by news reports of pesticide exposure in as far-reaching places as the United Kingdom, the Philippines, and Hawaii as well as reports from the World Health Organization and others, pesticide drift is a global problem. Residents of agricultural areas live in a toxic soup, exposed to innumerable, dangerous chemical pesticides on a daily basis. Formally trained experts and laypeople alike give us a wide range of compelling evidence to consider pesticide drift egregious, pervasive, and worthy of serious attention. We find ourselves shocked and disturbed by residents' stories of large-scale pesticide drift events, their tales of everyday pesticide exposure, scientists' admissions of how little we know about the thousands of pesticides that are widely used and how industry shapes regulatory practice, and by the plain fact that most of those chemicals are subject to no oversight and few restrictions.

In the past forty years, industry, the state, and mainstream activist groups have put considerable effort into addressing the environmental impacts of agricultural production—including those of toxic chemical pesticides. Various actors in the agricultural industry have invested in less toxic pest control technologies, supported pesticide applicator training programs, and participated in corporate sustainability reporting and certification initiatives. The environmental regulatory state has actively supported those industry efforts, and also implemented many regulatory restrictions on the use of some of the most toxic pesticides, streamlined the regulatory process for reduced-risk pesticides, and created many opportunities for public comment. The alternative agrifood movement has promoted organic agriculture and other less toxic pest management farming systems, and has helped to develop and maintain the institutions needed to support them. Actors throughout the world pursue similar approaches to environmental sustainability. What makes California a particularly fascinating case study is the fact that the extraordinary leadership and creativity demonstrated by its industry, regulators, and activists in addressing environmental problems have had little material impact on the pesticide drift problem itself. California thus casts into sharp relief the barriers to effective environmental problem solving.

One of my objectives in writing this book was to explain how a problem like pesticide drift came to be—how regular, relentless poisoning became both pervasive and invisible. Echoing the work of many other scholars to date, I have shown that toxic chemicals poison people's bodies in large part because the forces of capitalist competition compel businesses to exploit people and the environment in order to stay profitable, and because the state has failed to hold industry accountable for those harms. Yet I have also emphasized that exploitation and regulatory failure do not fully explain the problem of pesticide drift. I have argued that pesticide drift persists because the environmental sustainability efforts of the agricultural industry, the environmental regulatory state, and many relevant components of the alternative agrifood movement fail to account for many social issues that simultaneously exacerbate the problem as well as obscure it from view: material inequalities, social relations of oppression, a lack of participatory parity, and inadequate basic capabilities.

These actors' efforts to address pesticide drift are not a random collection but instead represent the ways in which predominant notions of justice shape the practices of industry, the state, and mainstream activist groups. This was my second objective in writing this book: to examine the underlying theories of justice that shape mainstream environmental politics, and critically interrogate their implications for public health and environmental problem solving. Throughout the book, I have identified the various roles that these predominant theories of justice—especially libertarianism and communitarianism—play in mainstream environmental politics. In some cases, these theories explicitly constitute the normative charge for particular programs, as is the case with the shift to market-based solutions in the regulatory arena. Similarly, industry actors often strategically (and incompletely) co-opt these notions of justice in order to discursively legitimize the alternatives to regulation that best serve their own interests. In other cases, actors unintentionally invoke and comply with libertarianism and/or communitarianism in order to achieve environmental gains through the path of least resistance, as is the case within many realms of the alternative agrifood movement.

What makes these ideological commitments, co-optations, and invocations significant is that although compelling, libertarianism and communitarianism are fundamentally unable to deal with problems like pesticide drift that require substantial government intervention and transcend any given community's good intentions. These limitations are obscured by the fact that social inequalities and oppressive social relations render the pesticide drift problem relatively invisible, thereby granting further legitimacy to libertarianism and communitarianism along with the policies and practices they endorse. In other words, both libertarian and communitarian notions of justice ignore—and frequently exacerbate—existing material inequalities, relations of oppression, disparities in different actors' abilities to participate in environmental decisions, and gaps in basic capabilities. They thus ignore and exacerbate the very factors that make today's most consequential environmental problems so intractable.

The debates over pesticide drift, then, provide an opportunity to critically reflect on the strengths and weaknesses of the particular visions of justice that govern mainstream environmental politics today as well as those that could possibly help address environmental inequalities more effectively. Pesticide drift activism is a reaction to both the material problem of pesticide drift and the problematic notions of justice that variously guide, legitimize, and/or are reinforced by the efforts of the crop protection industry, the pesticide regulatory state, and the alternative agrifood movement to address the problem. In contrast to the communitarian assumption that justice coheres at the local scale and the libertarian belief that justice requires unfettered market transactions, pesticide drift and other EJ activists assert that justice is a process of accounting for and combating environmental and social inequalities, the oppressive social relations that cause them, a lack of participatory parity in decision making on environmental and other social issues, and a lack of basic capabilities.

Accordingly, EJ requires a strong environmental regulatory state that accounts for inequality, oppression, lack of participatory parity, and inadequate capabilities, and proactively works with other state agencies and community groups to rectify these forms of injustice.

Pesticide drift is not, of course, the only environmental hazard or other social problem that pesticide drift activists face, nor is it the only one they politically confront. For these activists, though, pesticide drift is one especially illustrative way in which their lives are not adequately and fairly respected and protected. Pesticide drift is a powerful lens that they use to show the uncontrollability of agricultural pesticides and therefore advocate for health-protective environmental regulations. Pesticide drift activism reminds us that protecting consumers and workers from exposure to pesticides (however imperfectly that is actually done) only addresses part of the problem. By fixing our gaze on pesticide *drift*, activists demonstrate that ensuring a safe workplace for farmworkers—long seen as the gold standard in pesticide regulations—is not an effective way to protect the broader public from pesticide exposure. In fact, worker protection education and personal protective equipment can serve to legitimize and reinforce pesticide-intensive agricultural systems. The lens of pesticide drift emphasizes the unruly nature of chemicals in the environment and the need to reduce their presence in the broader landscape.

Pesticide drift activists have made numerous accomplishments in recent years. In their research on the EJ movement, Cole and Foster stress that grassroots organizing and coalition building are perhaps the two most crucial practices that EJ groups engage in. Pesticide drift activists devote considerable resources to these tasks and remain committed to a decentralized coalition of hundreds of community-based organizations across the United States. Despite charitable foundations' general reluctance to fund community organizing, pesticide drift activists continue to help existing and new community groups organize politically to fight pesticide drift in their local regions and at the state and national levels. These efforts have indisputably led to increased reporting of suspected pesticide exposures and improved visibility of the issue in the media as well as put greater political pressure on regulatory agencies to take health-protective actions. Pesticide drift activists have also played a lead role in forcing California regulatory agencies to reduce the contribution of agricultural pesticides to ground-level ozone (smog) and restrict the use of several highly toxic and drift-prone pesticides. In recent years, the Drift Catcher and Biodrift community-based participatory research programs have helped inspire community groups to organize, collect data, and join pesticide drift activism. The Drift Catcher and Biodrift projects have also started to gain legitimacy within state institutions that had initially regarded activist data with considerable skepticism: the California Department of Public Health invited pesticide drift activists to participate in its new large-scale biomonitoring project, and various U.S. EPA committees have referenced PAN's Drift Catcher data as "valid" evidence that indicates the need for increased regulatory attention to pesticide drift.[1]

Institutionalizing EJ

Given that pesticide drift activism focuses our gaze on the state, how might theories of justice help guide the task of strengthening and reforming the environmental regulatory apparatus? The time is ripe to be asking such questions, since the process of integrating EJ principles into the state's work has already begun.[2] The notion of institutionalizing EJ gained traction with President Bill Clinton's Executive Order on Environmental Justice (signed in 1994), the advent of the National Environmental Justice Advisory Council, and the establishment of the U.S. EPA's Office of Environmental Justice. The Executive Order directs federal agencies to incorporate EJ into their work and provides specific instructions for how they should do so.[3] Since that time, many U.S. states have passed EJ laws of their own.[4] Such policies do the important work of legitimizing the call for EJ and beginning the monumental task of operationalizing the EJ critique in a practical manner within state institutions.[5]

A number of observers, however, suggest that there are good reasons to be critical of these initial efforts to institutionalize EJ. Scholars and activists have observed that the U.S. EPA and legislators have generally interpreted EJ narrowly—as a requirement to crack down on intentionally discriminatory acts, "rather than a more proactive emphasis on the structuring of social and ecological relationships."[6] In addition to critiques from EJ activists and scholars, both the GAO and U.S. EPA's Office of Inspector General have criticized the U.S. EPA in recent years for generally failing to implement the Executive Order on Environmental Justice.[7] Activists and scholars railed against the Bush administration in 2005 for attempting to neuter the concept of EJ entirely; specifically, the EPA threatened to drop race "as a factor in identifying and prioritizing populations that may be disadvantaged by the agency's policies, asserting that all communities should be treated equally regardless of their race or socioeconomic status."[8]

Holifield found that U.S. EPA EJ programs purportedly designed to build political empowerment and economic self-sufficiency in marginalized communities actually served to "embed the neoliberal project more deeply in civil society," since they worked to "build trust," smooth over the gaps left by the neoliberal rollback, and otherwise neuter political anger, which in turn sustained rollback neoliberalism and removed it from political contestation.[9] The programs did address the need to boost capabilities in disadvantaged communities (namely, by increasing their access to grants), but they ignored processes and relations of oppression, did not reduce inequalities in risk, and did so little with their improvements to participation that Holifield characterized them as essentially a public relations effort for the agency. In reflecting on his conclusion that the EJ mandate appeared to have been operationalized in order to "define and manage the EJ community," Holifield and his colleagues recently argued that those trends in the U.S. EPA "threaten to blunt the more radical edges of the environmental justice frame."[10] As indicated in the first epigraph that opened this chapter, Obama administration EPA leaders have signaled that they will honor the pursuit of EJ in ways that the Bush administration certainly did not. Time will tell how meaningfully they have been able to follow through on this pledge.

In other words, the extent to which efforts to institutionalize EJ actually serve the definition of justice promoted by pesticide drift activists and the broader EJ movement depends precisely on how regulators put EJ concepts into practice. At this point in time, the track record suggests that there is considerable room for improvement. The task is certainly monumental; Cole and Foster characterize it as "turning the ocean liner."[11] Yet the notions of justice suggested by pesticide drift activists and the broader EJ movement provide a compass for guiding the institutionalization of EJ principles in the environmental regulatory arena. First and foremost, institutionalizing EJ means taking seriously inequality, oppression, a lack of participatory parity, and inadequate basic capabilities in all aspects of environmental regulatory practice. Environmental regulations must at the very least *account for* the effects of these forms of injustice in order to address the ways that they both deepen environmental problems and render them invisible within current regulatory practice. A full account of what this would look like is beyond the scope of this chapter, but a few suggestions can be made. The precautionary principle gives us a good place to start.

The Precautionary Principle

As discussed in chapter 4, the precautionary principle can serve as a recognized framework for helping to operationalize the EJ movement's theory of justice within the regulatory arena. In addition to the fact that regulatory science's ability to protect public health is already compromised by virtue of its reductionist analysis of singular chemicals, numerous scientific uncertainties, and various scientific norms, pesticide drift activism illustrates how oppressive social relations and the other forms of injustice further compromise risk assessment's ability to calculate risks as well as guide pesticide regulations. To draw on just a few of the many examples highlighted in this book: material inequalities prevent those who are most at risk of pesticide exposure from being able to escape or report suspected pesticide illness, or even (in the case of farmworkers) to adhere to existing workplace safety regulations, while industry uses its disproportionately large resources to fund and disseminate scientific research that supports its investments; regulatory science structurally oppresses agricultural community residents by sidelining lay, nonexpert knowledge and ways of speaking; existing public participation mechanisms do not meaningfully enable the public to participate until the terms of debate have already been set; and the lack of basic capabilities in many agricultural communities limits the reliability of the formal data on pesticide exposure that regulatory scientists use to reevaluate the safety of existing pesticides. If justice requires accounting for and fighting oppression, inequality, and inadequate participation and capabilities, then the precautionary principle supplies a guiding framework for institutionalizing EJ within regulatory practice. Precaution-based regulations are needed to account for both what we do know and what we can reasonably assume to be the case.

Numerous scholars have noted that EJ activists have just recently come to widely embrace the precautionary principle.[12] Phil Brown observes that the precautionary principle appeals to a wide range of EJ activists, helps to unite seemingly disparate groups, and offers EJ groups a framework that explains the social injustices that cause unequal burden.[13] The EJ movement in turn contributes to the precautionary approach. Through illuminating the causes of environmental inequalities, the EJ movement highlights social issues that further constrain scientific understandings of environmental problems like pesticide drift. In so doing, the EJ movement is helping to strengthen the argument for precaution-based regulatory reform.[14]

The development and dissemination of the precautionary principle in the United States have been inhibited by the fact that the environmental regulatory state has been oriented toward managing pollution rather than reducing it.[15] The bias toward pollution management within U.S. environmental regulatory practice stems from both material structures (notably, risk assessment) and cultural ones (such as regulators' widespread adherence to the risk and pesticide paradigms, and the long-standing bias toward productivism in U.S. political culture). The George W. Bush administration waged a particularly active, explicit, and "concerted campaign against the precautionary principle itself."[16]

Whiteside argues that misconceptions about what the precautionary principle means have hindered its development in the United States. Fearing that precaution-based regulations could lead to higher costs for businesses, industry has vociferously dismissed the precautionary principle as economically prohibitive and politically radical and lobbied against its implementation in the United States. It was for these reasons that the agrichemical industry aggressively (and successfully) worked to strip the concept from all discussions of EJ during the California EPA's deliberations on how to fulfill its new EJ legislative mandate.[17] California Assembly member Tom Barryhill (who is also a grower and staunch defender of the agricultural industry) articulated these concerns at a legislative hearing on methyl iodide in August 2009, illustrating the stridency with which many actors from the agricultural and crop protection industries lambast the principle:

Where is the concept of reasonableness rather than the emotional debates we see inserted into this process, and why do we seem every day to be moving closer to an extreme precautionary principle position? We're all concerned about long-term impact. But if we're so concerned to say, "Better safe than sorry, let's just not do it," we wouldn't be farming at all. There has to be a concept of reasonable risk.[18]

Certainly, precaution-based regulatory restrictions on specific agricultural chemicals would pose economic hardship for the businesses that manufacture those chemicals and, temporarily, the growers who use them (while they transition to less toxic alternative pest management regimes). As scholars like Whiteside and Brown point out, however, arguments like the one stated above by Barryhill contain several misrepresentations. First, they misrepresent the precautionary principle as an argument for prohibiting all chemical technologies instead of a call for a range of potential actions whose implementation would be contingent on the degree of scientific certainty and the seriousness of the potential risk. Second, such assertions misrepresent the relationship between precaution and the economy. Precaution-based regulations can promote the development and adoption of greener technologies, which prompt job growth in those new sectors. Environmental regulations are generally not correlated with job loss.[19]

Though the precautionary principle runs into these obstacles, its practicability is evidenced by the fact that other countries, particularly in Europe, have made progress in implementing it.[20] Many countries in Europe have banned chemicals deemed likely to be exceptionally toxic to human health, too difficult to control, and unnecessary given the availability of alternatives. The European Union's Registration, Evaluation, Authorisation, and Restriction of Chemicals initiative, made into law in 2007, further standardizes and operationalizes the precautionary principle in Europe. It requires industry to conduct certain levels of testing on chemicals (including those already in use), requires an alternatives assessment and cost-benefit analysis for each chemical in use, and places restrictions (consistent across the European Union) on the use of those deemed to be highly dangerous.[21]

Although officials in the United States have widely criticized the precautionary principle (even characterizing it, to quote one former senior administrator in the U.S. Office of Management and Budget, as a "mythical concept, perhaps like a unicorn"), scholars have shown that it has long been evident in U.S. environmental and public health law and policy, and thus is a reasonable guidepost for future reforms.[22] For example, all uncertainty factors in the risk assessment process constitute precaution-based practice, since they are health-protective recognitions of the limits of scientific knowledge. At the federal level, FIFRA's cost-benefit mandate provides the statutory basis for alternatives assessment in the U.S. EPA's pesticide regulatory process. While the EPA has historically interpreted the mandate to assess benefits narrowly, a broader conceptualization of alternative pest management methods in a pesticide's cost-benefit analysis could essentially operationalize alternatives assessment. (This would, though, require a revision of FIFRA's language that explicitly precludes alternatives assessment.)[23]

The California Environmental Quality Act directs DPR to eliminate hazardous pesticides as feasible alternatives become available and explicitly notes that "no action" (i.e., not registering a particular pesticide) must be recognized as a viable option.[24] Similarly, the California Food and Agriculture Code authorizes DPR to cancel any pesticide whose risks outweigh its benefits, or for which "there is a reasonable, effective, and practicable alternate material or procedure that is demonstrably less destructive to the environment."[25] California's Birth Defect Prevention Act of 1984 provides additional impetus for precaution-based pesticide regulatory reforms, as it was designed to eliminate the use of pesticides that cause birth defects and cancer. Federal and state laws already supply the legal basis for cumulative risk assessment in pesticide regulation as well. Notably, the federal FQPA of 1996 requires the EPA to assess the cumulative risks posed by pesticides that share a common mechanism of action.[26] Also, California's Health and Safety Code explicitly requires state officials to take cumulative risk into account when evaluating the safety of individual chemicals.[27]

The precautionary principle is not "mythical" or "extreme" but instead already manifest in many major institutions' guiding frameworks. The field of industrial hygiene, which largely guides the work of Occupational Safety and Health Administration staff and decision-making procedures, is a good case in point.[28] The industrial hygiene hierarchy of controls is consistent with the precautionary principle, as it prioritizes the elimination of hazard from the workplace (i.e., avoiding the use of hazardous pesticides, and replacing them with less toxic alternatives) and uses personal protective equipment requirements only as a last defense for protecting worker safety (i.e., when all other possible hazard avoidance and control strategies have been implemented).[29]

In other words, legislative and institutional groundwork have already been laid to put the EJ movement's theory of justice into action. Yet this will require an overarching, articulated precautionary principle, which as Whiteside shows, is missing in both the United States and the European Union, where forays into precaution have been ad hoc and discretionary. With the addition of a legally robust, guiding principle of precaution, the existing precaution-based statutes can be used much differently than they currently are—to help reduce pollution, protect public health, and pursue EJ.

Putting Precaution into Practice: The Case of Methyl Iodide

What might it look like to put the precautionary principle into practice? The case of methyl iodide—the new soil fumigant that was registered in 2008 by the U.S. EPA and in 2010 by California's DPR—provides some illustrative insights into this process. The case of methyl iodide also highlights that precaution-based regulations are already practiced, even with this economically valuable pesticide: regulatory agencies in the states of New York and Washington declined to register methyl iodide, citing potential "unreasonable adverse effects on human health" as their justifications for doing so.[30]

Although California and U.S. authorities have registered methyl iodide for use, the case is still open and thus appropriate for our evaluative purposes here—in the final days of 2010, pesticide drift activists sued DPR for registering the chemical and submitted 52,000 letters to Governor Jerry Brown asking him to revoke its registration. Additionally, the U.S. EPA has vowed to reconsider its decision in light of what transpires in California.[31]

Granted, some of the most glaring issues with regulatory decisions about methyl iodide have nothing to do with precaution: DPR officials' disregard for the health benchmarks set by their own scientists in the risk assessment is a notable instance, as is the use by the EPA and DPR of shoddy, industry-conducted scientific studies that violate numerous scientific conventions. Notwithstanding these regulatory failures, in this section I point out several of the major scientific controversies around methyl iodide that provide the basis for describing what it might look like to operationalize the precautionary principle in risk assessment and risk management. I draw in particular on written statements and public testimony from DPR and the external scientific review committee that DPR had commissioned to review its methyl iodide risk assessment.

One of the major points of contention over the registration of methyl iodide is that the dose-response portions of regulatory agencies' risk assessments suffer from several critical gaps in scientific data.[32] For example, methyl iodide is known to be a neurotoxicant (defined as a compound that disrupts the structure and function of the brain; laboratory and real-world exposure studies show that methyl iodide exposure leads to lasting neurological damage, including severe psychiatric and movement disorders that resemble Parkinson's disease). It is also known to be a developmental toxicant (meaning that it has increased effects on the fetus and young child). These facts indicate that methyl iodide is also presumably a developmental neurotoxicant (defined as a compound that increases the vulnerability of the developing brain to toxic insult, leading to neurobehavioral deficits such as affective disorders, learning disorders, autism spectrum disorder, and lowered IQ). No developmental neurotoxicity study, however, has ever been conducted on methyl iodide. Whereas the U.S. EPA and DPR assume that methyl iodide's risk of developmental neurotoxicity is zero, a precaution-based approach to risk assessment would make reasonable and health-protective guesses about methyl iodide based on what we already do know about it and structurally similar chemicals, and refrain from registering the chemical until an adequate developmental neurotoxicity study has been conducted by an independent laboratory. Scientific research committee members' testimony at a legislative hearing in June 2010 emphasizes this recommendation as well as the dangerous consequences of not doing so:

When we come across a compound that is known to be neurotoxic as well as developmentally toxic and an endocrine disruptor, it would seem prudent to err on the side of caution, demanding that the appropriate scientific testing be done in animals instead of going ahead and putting it into use, in which case the test animals will be the children of the state of California.

The scientific review committee also identified the problematic ways in which the exposure assessment portions of the EPA's and DPR's methyl iodide risk assessments are based on how pesticides would be used in ideal or so-called average scenarios, rather than how pesticides are actually used and experienced in the real world. For instance, the scientific review committee noted that regulatory agencies' estimates of worker exposures to methyl iodide unrealistically assume that workers work eight-hour days (whereas many work ten or more hours in a day), "strenuous" activity (with its higher breathing rates and thus higher exposure rates) accounts for only one of those hours, workers never remove their respirators for that entire time (even though they actually do need to do so in order to talk, drink, and eat), respirators form a perfect seal with each worker's face (even though dimples, scars, hair, and sweat will all invariably compromise those seals in actual use), and respirator filters are replaced daily.

In a hearing in June 2010 on DPR's proposed methyl iodide regulations, California Assembly member Bill Monning interrogated DPR director Warmerdam about these concerns that the scientific review committee raised. Her responses reveal one way that DPR dismisses such critiques:

Monning   You stated very clearly that MeI *can* be used safely. My real concern is *will* it be used safely. I think your "can" be used safely contemplates compliance with the mitigation proposals that your department has put forward that you said would be subject to EPA review. Do you build in any risk factor of the likelihood in the workplace of those mitigation factors not be[ing] fully or 100 percent complied with?

Warmerdam   We do *not* take into account those who disregard the law and do not comply. That is a violation and subject to enforcement. . . . It is difficult for us to regulate either stupidity, ignorance, or ignoring the law.

A few minutes later in the hearing, Warmerdam characterized a hypothetical worker's safety violation as "a bad personal decision on his or her part," which supervisors should address by "reassign[ing] an employee who is breaking the law." In other words, rejecting the assertion that the proposed regulations are unrealistic, she depicts applicators who fail to comply with the regulations as "bad apples" either flagrantly defying reasonable laws or too ignorant to be able to do their job. In addition to the fact that the proposed regulations contain numerous impossible expectations (e.g., that workers not remove their masks for eight full hours), Warmerdam's argument flies in the face of considerable health and safety research that has demonstrated that workers are likely to engage in dangerous activity when they feel they have little control over the safety of their work environments. Given farmworkers' low wages, legal status issues, language barriers, and the high competition for agricultural jobs, they are often neither sufficiently free nor informed to identify and contest pesticide safety violations. Referencing his past experience as an attorney, Monning recognized this reality and expressed concern about

workers terminated from a job for complaining about the employer not following safety regulations. I have represented those workers in the past and it creates a deterrent for people standing up and calling OSHA [Occupational Safety and Health Administration] or calling the ag commissioner because of fear of losing their job.

In contrast, by defining such problems as evidence of calculated recalcitrance or ignorance, and situating them squarely in the realm of compliance, Warmerdam thus dismisses the need to adjust the risk assessment's exposure evaluation.

In that same hearing, Warmerdam further defended DPR's exposure assessments by claiming that the state's regulatory apparatus contains multiple mechanisms for detecting any potentially inadequate pesticide regulations. First, she claimed that county agriculture commissioners have their "boots on the ground," implying that they will be able to observe pesticide illnesses and other evidence that DPR may have underestimated human exposures to methyl iodide. Warmerdam also contended that DPR's air-monitoring and periodic registration reevalution processes

give us a high level of confidence that, one, we will get the information we need to assure ourselves that the mitigation is appropriate and, secondly, if the information suggests otherwise, that we have an affirmative obligation to respond to what the information is telling us and make appropriate adjustments.

Yet as I have shown, CACs are widely biased in favor of the pesticide paradigm and frequently dismiss reports of pesticide exposure. Pesticide illnesses also are widely underreported and do not account for the delayed-onset health effects of pesticide exposure. Moreover, air-monitoring studies are expensive and rarely conducted, and relying on them to improve the risk assessments effectively positions residents as the guinea pigs of hasty regulatory decisions.

In contrast, a precaution-based approach would proactively account for and redress such data gaps and unrealistic assumptions. The scientific review committee members reiterated this argument throughout their written comments and oral testimony at various hearings, demonstrating that the precautionary principle is not radical but instead is actually considered reasonable practice among the nation's preeminent expert scientists:

We were very moved to tears when we heard from farmworkers at our hearing and they told us what the real world is like. The real world is not like what you hear in some labels that somebody promises they are going to adhere to. We are all grown-ups, and we all know whether or not people adhere to those kinds of rules.

If you are a risk assessor, it seems to me, you ought to be bound to estimate the exposures and the risks according to what is actually likely to happen, not what the theoretical label requirements say.

It is a mistake for those of us responsible for the health of Californians to be thinking about [what is] protective in a manner that is in ideal conditions. We need to treat it the way it is actually going to happen.

We live in the real world. We don't live in a world where labels guarantee success. . . . [W]e need to make sure that we have some confidence that there is going to be effective protection.[33]

Taking seriously the realities of social and ecological life in "the real world" would therefore require methyl iodide's risk assessors to incorporate precaution in their exposure assessments. In terms of worker exposures, this would include assuming that workers work long days (i.e., ten hours rather than eight), reducing the expected protections of respirators and other personal protective equipment (as the Occupational Safety and Health Administration already does), and increasing workers' expected breathing rate to account for the arduous nature of agricultural work. To make bystander exposure assessments more realistic, risk assessors should assume that people walk and play in the streets as well as on the sidewalks, and thus refrain from including such spaces in buffer zones. Risk assessors must also account for all possible routes of exposure (including groundwater contamination, which has hitherto been ignored).[34] Risk assessments should also include a precaution-based "environmental uncertainty factor" to account for the possibilities of wind gusts and animals that can break protective tarps, temperature inversions that prevent fumigants from dispersing quickly enough, and other elements of agricultural environments that often thwart the effectiveness of pesticide regulations—none of which are presently accounted for in regulatory agencies' exposure assessments of any pesticide. By assuming that such things might happen, a precaution-based approach would reduce the amount of methyl iodide that could be used in a given area in order to reduce potential bystander exposures. Additionally, the precautionary principle indicates the need for increased and independent environmental monitoring and laboratory research to fill in scientific data gaps—studies that could be funded through increasing registrant fees and/or implementing a federal pesticide sales tax.[35]

Despite these practical improvements to risk assessment, its predictive capacity is still basically limited by its attention to single chemicals in isolation. Proponents of the precautionary principle accordingly call for more fundamentally restructuring risk assessment in two key ways. First, a health-protective, precaution-based risk assessor would subject methyl iodide to a *cumulative risk assessment* to try to account for the facts that agricultural pesticides are often used in mixtures, and that workers and residents are exposed to various combinations of such mixtures. Risk assessors could evaluate the risks associated with exposure to the mixture in which methyl iodide is actually used (which contain anywhere from 2 to 75 percent chloropicrin, a potently carcinogenic fumigant).[36] Furthermore, a cumulative risk assessment would limit the amount of methyl iodide that could be used in a broader geographic space at any given point. Although the proposed regulations limit the size of a field that can be fumigated at a time and prohibit fumigations on adjacent fields in a single day, they do not try to limit the number of applications in a larger region.

Second, because risk assessment's focus on single chemicals makes it structurally unable to account for necessity, a health-protective, precaution-based risk assessor would subject methyl iodide to an alternatives assessment to determine whether it is even needed in light of the alternatives. As is the case with cumulative risk, alternatives can be interpreted in many different ways. Among industry actors and policymakers, methyl iodide has been widely recognized as toxic, but characterized as the only viable "drop-in replacement" for methyl bromide. This is so clearly the hegemonic interpretation of alternatives that Arysta LifeScience, the manufacturer of methyl iodide, won the U.S. EPA's Stratospheric Ozone Protection Award in 2009 and several other environmental awards for having developed a replacement for methyl bromide.[37] In contrast, pesticide drift activists, other environmental activists, organic farmers, and many scientists advocate a broader conceptualization of alternatives that includes not just instantly substitutable chemical inputs but also alternative cropping and pest management regimes that avoid fumigant use entirely.[38]

A precaution-based assessment of methyl iodide would therefore consider the need for methyl iodide in light of all its alternatives—not just methyl bromide, but also the other existing soil fumigants and alternative agricultural pest management regimes that avoid the use of any highly toxic soil fumigant. That said, nonfumigant alternatives are often not as effective at controlling pests and in some cases require rotating in less profitable crops to disrupt the pest patterns that accumulate with regular planting of the same crop. Advocates of alternatives assessment must recognize that both of these limitations can pose economic hardships for growers who have relied on methyl bromide in the past. Thus, to make some alternatives to methyl iodide realistic, the alternatives assessment might need to include a provision for directing some federal subsidy funds to marginalized farmers who transition to less toxic, nonfumigant methods of pest management.

Rectifying Environmental Injustices

While regulatory agencies can use the precautionary principle as a framework for accounting for the injustices that compromise existing environmental regulatory process and outcomes, the EJ movement's theory of justice also requires that the regulatory state play a role in actively *rectifying* those injustices. For example, the U.S. EPA and the California DPR have already started the process of confronting the lack of participatory parity by designing and implementing numerous public participation mechanisms that facilitate the participation of hitherto-marginalized groups. Note that the worker and resident testimony at public hearings on methyl iodide demonstrated to the scientific review committee many of the unrealistic assumptions frequently made in regulatory agencies' risk assessments. While regulatory agencies' public participation efforts are commendable, EJ requires that groups are allowed to not only speak about but also shape the material, environmental outcomes. To conclude her analysis of EJ activism in Louisiana's chemical corridor, Barbara Allen emphasizes the need for more inclusive, democratic regulatory process. She argues that this requires equal access to usable information for all stakeholders; recognition of and respect for lay observations and experiences; ongoing and regular opportunities for citizen input about community development (i.e., throughout the entire process in which potentially harmful facilities are evaluated, and extending to a wide range of development issues including but extending beyond siting toxic facilities); the ability of lay input to influence environmental change; and the establishment of opportunities to participate not just at the local level but also at higher-scale (state, federal, and global) levels of decision making.[39]

Although pesticide drift activist leaders worry that institutionalized public participation mechanisms can disempower participants by keeping them in meetings rather than out doing grassroots organizing, they also concede that there are some benefits. Numerous activists stressed to me in interviews that the public participation opportunities can serve as crucial moments for teaching communities how to mobilize, and enrolling their friends and neighbors into the cause. As one leader reflected, public participation mechanisms are "all parts of building and showing power." Having the chance to tell their story "and the perception of being listened to" is an important release, especially for new activists. As one activist stated about her participation in DPR's EJ meetings, "There is something empowering about this." This leader then reflected that public participation can also be a significant part of teaching people how slowly change happens, and given this, that they may need to adjust their expectations.

In addition to participation, institutionalizing justice can mean that environmental regulatory agencies actively confront oppression, such as through taking residents' concerns seriously, valorizing the goal of reducing pollution, and actively stopping the disparagement of people who do not adhere to the pesticide paradigm. The environmental regulatory state can actively fight the injustice of inequality through institutionally curbing industry's special access to regulatory decision-making processes, and continuing to actively solicit and incorporate the participation of groups that are marginalized yet highly affected by regulatory decisions. At a minimum, regulatory agencies can battle inequality simply by more fully enforcing existing laws—holding industry accountable for regulatory violations.

That said, the pursuit of EJ must extend beyond the pesticide regulatory state. Many other federal environmental and agricultural programs could be adjusted to help industry and pesticide regulators make this transition. Public university research and outreach priorities should continue to be redirected to explicitly reduce the environmental impacts of agriculture. With appropriate leadership from Congress during the Farm Bill deliberations, the USDA could commit a substantial portion of the federal agricultural budget allocations to research into alternative pest management practices, the development of distribution and marketing systems that support growers who practice such methods, and practical training and income assistance to support growers transitioning to less toxic pest management.

Combating social injustices also requires granting farmworkers the same rights as other workers; boosting labor rights for all workers; implementing meaningful federal immigration reform that honors immigrants, greatly expands legal immigration opportunities, and provides pathways to citizenship for unauthorized workers; closing corporate tax loopholes; and passing truly comprehensive health care reform. This is, admittedly, a tall order, and one that glosses over extraordinary complexities. The point is that environmental problems are inherently social ones. To achieve effective, just solutions to today's environmental problems, we must account for social injustices—inequalities, oppression, a lack of participatory parity, and inadequate basic capabilities—and actively work to rectify them.

This is the EJ movement's vision of justice in action, and part of what makes it so important is that it is qualitatively distinct from the ideas of justice that shape mainstream environmental politics today. Those predominant conceptions of justice—utilitarianism, libertarianism, and communitarianism—widely shape mainstream environmental politics but do not help us deal with environmental inequalities like pesticide drift. Instead, they ignore those inequalities, reinforce them, and render them more difficult to see. Environmental justice thus requires that we not only combat malfeasance, ineptitude, and negligence, but also that we think carefully about the multiple, and conflicting, notions of justice at work around us every day.

Notes

Chapter 1

1. Gelobter 2002, xiv, xiii.

2. Just to be clear, in this book I am concerned with *political* justice rather than the *legal* justice of the courts.

3. For excellent overviews of the U.S. environmental movement, see Gottlieb 2001, 2005. For literature on the EJ movement, see Agyeman, Bullard, and Evans 2003; Bryant 1995; Bryant and Mohai 1992; Bullard 1990, 1993, 2005; Camacho 1998; Cole and Foster 2001; Faber 1998; Faber and McCarthy 2003; Gottlieb 2001; Hofrichter 1993; Pellow and Brulle 2005; Pulido 1996a, 1996b; Sandler and Pezzullo 2007; Szasz 1994.

4. United Nations Environment Program 2004, 7.

5. For example, Murray and his colleagues (2002) estimate that 98 percent of the pesticide poisonings in Central America are not reported.

6. For literature on the role of chemical pesticides and other industry innovations in helping to shape the development of modern agricultural food systems, see Fitzgerald 2003; Goodman, Sorj, and Wilkinson 1987; Henderson 1999; Stoll 1998; Walker 2004. For literature on environmental sustainability efforts in U.S. agriculture, see Allen 2004; Allen and Sachs 1993; Bell 2004; Guthman 2004; Hassenein 1999; Warner 2007.

7. For overviews of agrifood activism in California, see Allen 2004; Guthman 2004; Gottlieb 2001. For historical accounts of specific periods of pesticide activism, see Pulido 1996b; Nash 2006.

8. The information sources include my personal interviews with victims, activists, and agency officials, DPR records (PISP 1999, 5, 10), and newspaper articles (including Maxwell 1999, 2000; Olvera 1999a, 1999b; Stapleton, 2003).

9. The table was constructed from data in DPR's PISP annual reports, available at <http://www.cdpr.ca.gov/docs/whs/pisp.htm>, and the California Pesticide Illness Query database, available at <http://apps.cdpr.ca.gov/calpiq>. The "Number of people affected" include those deemed "possibly," "probably," or "definitely"

affected by DPR. Note that DPR's Pesticide Illness Surveillance Program (PISP) data were only available through 2007 at the time that this book was written.

10. The information sources include my personal interviews with activists and agency officials; DPR records (PISP 2000, 9–10); and newspaper articles and opinion pieces (see, for example, Alvarez 2000a, 2000b; Corley 2000).

11. The information sources include my personal interviews with activists and agency officials, and DPR records (PISP 2002, 10–11).

12. Testimony at special legislative hearing on methyl iodide in Sacramento (DPR 2009f).

13. The information sources include my personal interviews with activists and agency officials, DPR records (PISP 2003, 9–10), and newspaper articles (see, for example, Hsu 2003a, 2003b).

14. Local officials consistently gloss over fact that the residents' initial reports of illness were ignored. As the local county agriculture commissioner stated in an interview with me, "So the good part was that we learned from Earlimart and Arvin, and were able to enact that response plan in Lamont, and it actually worked very well for getting information to the people who were out there." This claim was inexplicably supported by a grand jury investigation of the incident response (Pesticide Drift 2004).

15. The primary initial groups brought together in 1995 with the help of Pesticide Watch included residents from Warner Springs (San Diego County), Lompoc (Santa Barbara County), Castroville (Monterey County), Davenport (Santa Cruz County), and Santa Rosa (Sonoma County). Central Valley activism gained considerable momentum following the Earlimart and Arvin incidents.

16. As of January 1, 2010, DPR (2010a) reports that there are 983 pesticide active ingredients and 13,422 product formulations actively registered for use in California.

17. I draw here on Luke Cole and Sheila Foster's (2001) characterization of the EJ movement's roots. While emphasizing antitoxics activism and the civil rights movement as the main tributaries to the EJ movement, they also note that other key players are the labor movement, Native American struggles, and some academics.

18. Cole and Foster (2001, 29) explain that what is now considered the "traditional environmental movement" had its roots in the civil rights movement and anti-Vietnam activism, but generally dropped those social justice orientations in favor of a focus on "legal and scientific approaches to environmental problems": "litigation, lobbying, and technical evaluation" (see also Gottlieb 2005). This shift directly reduced many forms of pollution, but it excluded people seen as nonexperts and often intentionally sidelined questions of social justice.

19. Major explanations have pivoted around corporate consolidation as well as concentration in the farm input, processing, and retail sectors (Bonanno et al. 1994; Friedland et al. 1991; Magdoff, Foster, and Buttel 2000; McMichael 1994), university research and extension priorities and practices (Busch and Lacy 1983; Henke 2008; Hightower 1973; Kleinman 2003; Krimsky 2003; Vallas and

Kleinman 2008; Warner 2007), industry practices (Busch et al. 1991; DuPuis 2002; FitzSimmons 1986; Friedland, Barton, and Thomas 1981; Goodman, Sorj, and Wilkinson 1987; Henderson 1999; Kloppenburg 2005; Stoll 1998; Thomas 1985; Wells 1996), agricultural policies (Cochrane 1979; Lobao and Meyer 2001; Winders 2009; Ray, De La Torre Ugarte, and Tiller 2003), and labor and immigration policies (Daniel 1981; Galarza 1964; McWilliams [1939] 1999; Mitchell 1996). Academic and popular presses have published a wide range of books in the past few years that point to the ways in which aspects of popular food culture interact with industry marketing and restructuring practices to contribute to agrienvironmental pollution, public health problems, the consolidation of corporate power, and various forms of social inequalities. Those written for mass audiences introduce (however simplistically) key social issues in food and agriculture to the broader public in ways unseen since Upton Sinclair's work a century ago (notable in this regard are Michael Pollan's *Omnivore's Dilemma* [2006] and *In Defense of Food* [2008], Marion Nestle's *Food Politics* [2002], and Eric Schlosser's *Fast Food Nation* [2001]). In a similar vein, a handful of recent books from the interdisciplinary scholarly field of agrifood studies (including Mark Winne's *Closing the Food Gap* [2008], Melanie DuPuis's *Nature's Perfect Food* [2002], Suzanne Freidberg's *French Beans and Food Scares* [2004] and *Fresh* [2009], Wynne Wright and Gerad Middendorf's *The Fight Over Food* [2008], Clare Hinrichs and Thomas Lyson's *Remaking the North American Food System* [2007], and Lyson's *Civic Agriculture* [2004]) take on these same issues for a more academic audience.

20. For historical accounts of the pursuit of sustainable agriculture, see Allen 1993, 2004; Guthman 2004. For a few examples of visions of sustainable food systems, see Kloppenburg, Henrickson, and Stevenson 1996; Lyson 2004; Pollan 2006. On pragmatic institutional designs for promoting sustainable farming in the United States, see Bell 2004; Hassenein 1999; Lyson, Stevenson, and Welsh 2008; Warner 2007. For a few books emphasizing shopping and other lifestyle directives for consumers, see Lappé and Lappé 2002; Nestle 2006; Pollan 2006, 2008; Singer and Mason 2006.

21. An important subset of such work focuses on antitoxics activism. See, for example, Allen 2003; Brown 2007; Brown and Mikkelson 1990; Cole and Foster 2001; Corburn 2005; Sze 2007.

22. See Capek 1993; Benford 2005; Taylor 2000; Kurtz 2003. Aligning with what is now regarded as classic social movement framing literature (Snow et al. 1986; Snow and Benford 1988, 1992; Benford and Snow 2000), Stella Capek's influential article (1993, 5) emphasized that the EJ frame "has been fashioned simultaneously from the bottom up (local grass-roots groups discovering a pattern to their grievances) and from the top down (national organizations conveying the term to local groups)."

23. In this way, scholarship on antitoxics activism overlaps with the sociology of scientific knowledge as well as science and technology studies fields, which interrogate the social construction of expertise and knowledge validity.

24. See, for example, Cronon 1990, 1998; Guha 1998; Kosek 2004; Merchant 2003; White 1995; Williams 1973, 1980.

25. Most recently, academic observers of the EJ movement have started the important task of evaluating the EJ movement itself in constructive but critical terms. See Pellow and Brulle 2005; Pulido 1996a; Sandler and Pezzullo 2007.

26. See Dobson 1998, 2003. My discussion of political philosophy in this section (and throughout the book) obviously oversimplifies a tremendous literature and obscures its nuances, details, weaknesses, and internal debates. For an excellent introduction to and overview of this literature, I recommend Will Kymlicka's *Contemporary Political Philosophy: An Introduction* (2002). My oversimplification here is deliberate, since my objective is to distill the fundamental concepts of political theory (which exist in spite of all of the elaborate scholarly debate) so that they are accessible to a wider audience in the analysis of an empirical case study.

27. Young 1990.

28. Young invokes an explicitly Foucauldian conception of power, emphasizing that oppressive power is often productive and widely disseminated: "When power is understood as 'productive,' as a function of dynamic processes of interaction within regulated cultural and decisionmaking situations, then it is possible to say that many widely dispersed persons are agents of power without 'having' it, or even being privileged" (Young 1990, 33).

29. For example, see Sen 1993, Nussbaum 2005.

30. See Cole and Foster 2001; Hunold and Young 1998; Young 1983; Shrader-Frechette 2002; Lake 1996; Figueroa 2003; Pellow 2000.

31. Schlosberg 2007. Schlosberg points out that considerable EJ scholarship focuses on the distribution of environmental hazards, whereas most EJ activists' concerns are much broader than distribution per se.

32. I depart slightly from Schlosberg on one point. He argues that "the capabilities approach can be seen as fully incorporating recognition and distribution in a broad theory of justice" (2007, 34), and I think this overstates the case. In my opinion, the capabilities approach adds a fourth important element of justice to what Nancy Fraser (2000) calls a "trivalent" conception of human justice (distribution, recognition, and participation).

33. Schlosberg 2007, 51.

34. Egalitarians emphasize that the state has a responsibility to equalize people's circumstances in cases where inequalities stem from chance, or unequal natural or social endowments. Such a theory of justice lends credence to progressive tax structures, welfare programs, and other redistributive political institutions that help the poor and disadvantaged.

35. Harvey 1996, 375.

36. Ibid., 397.

37. Cronon 1992, 1375.

38. Jones 1998, 27–28.

39. Foucault (1978) 1990, 101.

40. Cowan 2005, 2.

41. Bullard 1993; Kuletz 1998.

42. Young 1990, 5.

Chapter 2

1. See Busch and Lacy 1986; Busch et al. 1991; DuPuis 2002; Fitzgerald 2003; FitzSimmons 1986; Goodman, Sorj, and Wilkinson 1987; Henderson 1999; Henke 2008; Hightower 1973; Jelinik 1979; Kleinman 2003; Stoll 1998; Walker 2004.

2. Guthman 2004.

3. This phenomenon is often referred to as the "cost-price squeeze" within agri-food studies literature.

4. Another increasingly utilized strategy by which growers maintain profits in the face of increasing costs and declining prices is adding value to their goods and selling in specialty markets. This approach—what Julie Guthman (2004, 66–67) calls "valorization"—is the principle that enables some growers to receive higher prices for labeling their products as organic, local, or fair trade.

5. See DPR 2006c; U.S. Department of Food and Agriculture 2002; U.S. EPA 2004, tables 3.5 and 4.2.

6. Examples include organic agriculture, integrated pest management, and agro-ecology. See Altieri 1987; Gliessman 2007; Benbrook 1996.

7. Bosso 1987, 28.

8. DuPuis 2002. See also Henke 2008; Stoll 1998; Russell 2001; Bosso 1987; Fitzgerald 2003. These principles, technologies, and rhetoric were exported throughout the developing world in the mid-twentieth century as a form of agri-cultural and political development known as the Green Revolution.

9. Villarejo et al. 2000, 8; Aguirre International 2005, 5.

10. Aguirre International 2005, 15, 26; Villarejo et al. 2000.

11. McWilliams (1939) 1999; Mitchell 1996, 2007.

12. Mitchell 2007, 567.

13. Nevins 2002, 2005, 2008; Purcell and Nevins 2005; Massey 2002; Andreas 2000.

14. Cornelius 2001; Department of Homeland Security 2009.

15. GAO 2006. See also Andreas 2001; Eschbach et al. 1999; Mitchell 2001, 2007; Nevins 2008; Cornelius 2001. Several immigrant farmworkers report that now some immigrant smugglers (coyotes) prefer to wait for a snowstorm before guiding unauthorized immigrants through mountain zones, as those are the cir-cumstances during which the area is least likely to be monitored by the U.S. Border Patrol agents. Such practices indicate that border militarization continues to make border crossing more physically dangerous for unauthorized immigrants, at the same time that opportunities for legal migration have not increased and drivers of immigration have continued unabated.

16. Varsanyi 2008a, 2008b; Winders 2007; Coleman 2007.

17. Stephen 2004; see also Coleman 2007.

18. Holmes 2007. For historical accounts, see McWilliams (1939) 1999; Mitchell 1996.

19. Williams 1973; Mitchell 1996.

20. Chavez 2001.

21. U.S. Census Bureau data from 2000 indicate that 17 to 19 percent of households in Earlimart, Arvin, and Lamont report incomes of less than ten thousand dollars per year, and that 88 to 90 percent of the residents in these towns are Hispanic/Latino.

22. California 2010.

23. Employers regularly complain about labor shortages. Labor shortages, however, tend to be regional and temporary, and farm labor markets are generally flooded.

24. Kegley, Katten, and Moses 2003, 7.

25. Akesson and Yates 1964; Kegley, Katten, and Moses 2003.

26. A note about metam sodium and metam potassium: These two pesticides' primary breakdown product (methyl isothiocyanate [MITC]) is actually what causes illness in exposed individuals. Another MITC-generating pesticide (dazomet) is used in California, but only in relatively small amounts, so for simplicity I do not mention it in this book. Although "MITC-generating compounds" would be a more accurate term to use here, for clarity I use the pesticide names (i.e., metam sodium and metam potassium).

27. Becker et al. 2005.

28. Pesticide use here is measured in pounds of active ingredients; see DPR 2008c. In contrast, fumigants account for 10 percent of all agricultural pesticides used in the United States. Fumigants appear lower on the top pesticide rankings nationally because farmers outside of California rely more on herbicides than on fumigants; the most recent (2001) nationwide data show fumigants at number three (metam sodium), number seven (methyl bromide), number eight (1,3-dichloropropene), and number eighteen (chloropicrin) for conventional pesticides. See U.S. EPA 2009g; Becker et al. 2005.

29. DPR (2010d) determined that chloropicrin is a carcinogen and causes chronic respiratory damage in its review and determination of the pesticide as a TAC. The neurotoxic effects of methyl bromide are elaborated in a recent review by the U.S. Environmental Protection Agency (U.S. EPA 2008c).

30. Indeed, David Pimentel and Lois Levitan (1986) argue that, due to drift and other factors, as little as 0.1 percent of all pesticides applied effectively reach their target pest.

31. DPR 2009i.

32. California DPR's PISP system annual summary reports. I included those cases that stemmed from agricultural pesticide applications and were classified by DPR staff members as "possible," "probable," and "definite" (which excludes PISP cases that are classified as "asymptomatic," "unlikely," or "lacking sufficient

information." From 1998 to 2007, the total number of reported agricultural pesticide illness cases was 4,214, and the total number of those that were attributed to drift was 2,617.

33. For these data, I used DPR's new online California Pesticide Illness Query (CalPIQ) database. Of the 2,567 pesticide illness cases attributed to agricultural pesticide drift from 1998 to 2007 in the CalPIQ database, 1,396 were associated at least in part with a fumigant and 518 with a major organophosphate (chlorpyrifos, diazinon, dimethoate, disulfoton, malathion, methamidophos, or oxydemeton-methyl). Note that CalPIQ data differ slightly from those of the annual report summaries that I used for figure 2.4 (i.e., CalPIQ reports 2,567 pesticide illness cases attributed to agricultural pesticide drift from 1998 to 2007, whereas the annual summaries indicate 2,617 for that period), because some of the data have been revised since the annual reports were originally published. Given that the difference is so small, I draw on both sources of information, as each one provides certain details that the other does not.

34. National Institute for Occupational Safety and Health 2009.

35. Moses et al. 1993, 940.

36. The EPA does estimate that ten to twenty thousand physician-diagnosed pesticide illnesses and injuries occur in farmwork alone each year. Nevertheless, the agency admits that "its estimate represents significant underreporting and that no comprehensive national data are available on the extent of pesticide illnesses" (GAO 2001, 1). For critiques of U.S. EPA pesticide illness data, see Levine 2007, 40. It is worth noting that residents in agricultural communities have reported pesticide exposures and illnesses since at least the 1950s (Nash 2006, 257n75; Stoll 1998).

37. See Moses 1993; Moses et al. 1993; Reeves, Katten, and Guzmán 2002.

38. For a description of the basic components of the process, see DPR 2009i. For critiques, see Moses et al. 1993; Levine 2007.

39. DPR also categorizes a few cases as asymptomatic and "not applicable." Definitions of the categories are discussed in DPR 2002b, 55; 2009i.

40. CPR 2009b.

41. Quoted in Vasquez 2005. Don Villarejo and Stephen McCurdy (2008, 143) surveyed farmworkers and found that the "number of workers who reported having had direct contact with pesticides during the twelve months prior to the interview was not small: 12% of men and 7% of women reported having been sprayed or drifted upon." This study's results may differ from that of CPR (2009b), because of the different time frames specified in the questions (i.e., "during the [past] twelve months" for the former versus "ever" for the latter).

42. DPR 2002a. For the Toxc Air Contaminants reports, see DPR 2009j.

43. PAN has critically evaluated ARB/DPR air-monitoring results for soil fumigants, chlorpyrifos, and diazinon (Kegley, Katten, and Moses 2003), chlorpyrifos (PAN 2006), endosulfan and diazinon (PAN 2007), and chlorothalonil (PAN 2008).

44. Dansereau 2009; see section 2. See note 26 above regarding MITC.

45. This air-monitoring study is detailed in Meuter 2008.

46. Tupper, Kegley, and Hill 2008; Farm Worker Pesticide Project 2006; PAN 2006. Other Drift Catcher studies have found problematic levels of other pesticides in additional locations throughout the United States (see PAN 2009b).

47. CPR 2007. The U.S. Center for Disease Control and Prevention periodically conducts biomonitoring on a large sample of the U.S. population to test for human exposure to a wide variety of chemicals. Though regularly demonstrating widespread exposure to many pesticides and other chemicals, such studies are of limited utility to pesticide regulators, since the data do not indicate the source of contamination, distance between the individual and the sources of contamination, or the route of exposure (i.e., dietary, dermal, or inhalation).

48. Thornton 2000, 30–31.

49. Pearce and Mackenzie 1999; Charizopoulos and Papadopoulou-Mourkidou 1999; Glotfelty, Seiber, and Liljedahl 1987; Domagalski et al. 2000; Dubrovsky et al. 1998; Schuette et al. 2002; Heavner 1999.

50. Zamora et al. 2003.

51. LeNoir et al. 1999. See also Aston and Seiber 1997; Zabik and Seiber 1991.

52. Author's analysis of ARB and DPR pesticide air-monitoring data; DPR 2009a, 2009d, 2009k; Baker et al. 1996.

53. DPR 2008b.

54. Exposure studies are of three types: dietary, dermal, and inhalation. To assess dietary exposure, risk assessors examine studies of pesticide residues on food. To assess dermal exposure, risk assessors analyze data from worker exposure monitoring studies (to determine the quantity transferred from crops to a worker's body, which is assessed for various tasks) and laboratory studies (to determine the chemical's transfer rate into the body). To assess inhalation exposure, risk assessors look at air-monitoring studies.

55. For a recent review of debates about risk assessment, see NRC 2009.

56. In this section, I draw in particular on critiques in Brown 2007; Nash 2006; O'Brien 2000; Thornton 2000.

57. Wing 2000, 34. Similarly, Joe Thornton (2000) compares risk assessment's narrow, mechanistic view of causality with a more dynamic, complex, ecological view of natural systems.

58. Porter, Jaeger, and Carlson 1999; Heindel et al. 1994; Boyd, Weiler, and Porter 1990, as cited in Lee et al. 2002, 1182.

59. Kegley, Katten, and Moses 2003, 41.

60. Marquardt, Cox, and Knight 1998.

61. California's Office of Environmental Health Hazard Assessment made this point in its critical evaluation of DPR's risk assessment of metam sodium (see Fan 2002, 5).

62. Linda Nash's recent book, *Inescapable Ecologies*, helps to explain the long-standing lack of research on environmental contributions to human disease. Nash

(2006) shows that predominant trends in public health research sidelined environmental explanations of health problems, despite the fact that researchers have expressed concern about the health effects of chronic exposures to low levels of pesticides since at least the 1930s (see also Stoll 1998, 120).

63. Colborn, Dumanoski, and Myers 1997; Langston 2010.

64. As many critical scholars of science have noted, this scientific convention of minimizing "Type I errors" (false positives) and thus potentially unnecessary regulations in favor of "Type II errors" (false negatives) elevates industry's needs over those of public health and the environment (see, for example, Raffensperger and Tickner 1999, 3).

65. Castorina and Woodruff 2003. For an excellent historical discussion of the cultural origins and limitations of toxicology and epidemiology, see Nash 2006, chapter 5.

66. Montague 2000, ix.

67. Lee et al. 2002, 1179.

68. Author's own interviews; Kegley, Katten, and Moses 2003.

69. Brown and Mikkelson 1990; Brown 2007.

70. Arguello et al. 2005, 4.

71. For academic analyses of epidemiological research investigating the relationships between pesticides and these health outcomes, see Sanborn et al. 2004; Moses et al. 1993; O'Malley 2004. See also Guillette et al. 1998; Roberts et al. 2007.

72. Davidson 2004; Davidson, Shaffer, and Jennings 2002; Hayes et al. 2002; Shah 2010.

73. For particularly thoughtful critiques of epidemiology, see Brown 2007; Nash 2006; O'Brien 2000; Thornton 2000, 106–113.

Chapter 3

1. Fitzgerald 2003; Stoll 1998; Walker 2004; Jelinik 1979.

2. Stoll 1998.

3. McWilliams (1939) 1999; Mitchell 1996, 2007.

4. Stoll 1998.

5. Ibid., 114.

6. For the first major (and particularly scathing) critique of land grant university research priorities, though principally concerned with the impacts of agricultural technologies on farm structure (notably, the survival of small-scale farms), see Hightower 1973. For more recent critiques of public university research, see Busch and Lacy 1983; Henke 2008; Kleinman 2003; Warner 2007.

7. Russell 2001.

8. DataMonitor 2008. Note that the overall sales figures were relatively stagnant from the mid-1980s until they began increasing again in 2004. The period of

stagnant sales largely reflected greater sales of less expensive, generic pesticides—that is, not a declining use of pesticides (see Dinham 2005).

9. Robbins 2007.

10. Dinham 2005; ETC Group 2005, 6.

11. Perrow 1984; International Coalition for Justice in Bhopal 2011.

12. De Guzman 2009.

13. Kloppenburg 2005.

14. National Coalition on Drift Minimization 2009.

15. DJSI 2009a. Bayer, Dow, Syngenta, and BASF were named to the DJSI (World) in 2008–2009.

16. According to the Responsible Care (2009) Web site, "Responsible Care is a commitment, signed by a chemical company's Chief Executive Officer (or equivalent in that country) and carried out by all employees, to continuous improvement in health, safety and environmental performance, and to openness and transparency with stakeholders. It helps companies improve performance by identifying and spreading good management practices, and promotes mutual support between companies and associations through experience sharing and peer pressure."

17. See DJSI 2009b.

18. Keystone Alliance 2009, 6.

19. King and Lenox 2000.

20. Hedberg and von Malmborg 2003; Rees 1997; King and Lenox 2000.

21. Rees 1997.

22. Ibid., 508.

23. Bartley 2007.

24. King and Lenox 2000.

25. Benbrook 2009.

26. Sulfuryl fluoride is a particularly potent greenhouse gas and is being tested for agricultural uses. In 2007, fifty-four scientists (most of whom are members of the National Academy of Sciences, and including five Nobel Laureates in Chemistry) urged now-former U.S. EPA Administrator Stephen Johnson not to register methyl iodide as a soil fumigant because of the high risks it poses to human health and the environment (Bergman et al. 2007). I discuss methyl iodide in detail in chapters 4 and 6.

27. Dinham 2005; Galt 2008; Weir and Schapiro 1981.

28. Robbins 2007.

29. Dinham 2005.

30. According to the self-reporting of thirty-seven agricultural publications, Syngenta, Monsanto, and Bayer spent $30 to $35 million annually on crop chemical advertisements in recent years (Panousis 2008).

31. Kroma and Flora 2003.

32. Robbins 2007.

33. Gennaro and Tomatis 2005; Mullenix 2005; Egilman and Billings 2005.

34. Patel, Torres, and Rosset 2005.

35. Stauber and Rampton 1995; Rampton and Stauber 2002. See also the Center for Media and Democracy 2011.

36. Tyrone Hayes's research on the herbicide atrazine and the manufacturer's efforts to suppress his findings is a damning case in point. See Hayes 2004; Land Stewardship Project and PAN 2010.

37. Kleinman 2003, 44. For a critical examination of industry-funded university research, see Washburn 2005; Krimsky 2003.

38. Kjaergard and Als-Nielsen 2002; Als-Nielsen et al. 2003; Krimsky 2003.

39. For example, Monsanto reported spending $8.83 million on lobbying in 2008 alone (Center for Responsive Politics 2009). The Center for Responsive Politics provides information on federal lobbying expenditures, and California's Secretary of State supplies information on reported state-level lobbying expenditures in California (California Secretary of State 2009). That said, formal lobbying expenditures represent only a fraction of the actual funds spent on lobbying-like activities. For a discussion of some of the latest trends in "not lobbying," see Delaney 2010.

40. Center for Responsive Politics 2009.

41. Spitzer n.d. For a collection of particularly egregious cases of the revolving door in action, see Source Watch 2009. See also Janine Wedel's discussion (2009) of the "evolving door"—a term she uses to characterize the less shocking but highly consequential movement of people between industry and state agencies.

42. Jacobson 2005.

43. Coalition for Chemical Safety 2009.

44. PAN 2009e.

45. CropLife 2008. This figure includes sales of pesticides, fertilizers, seeds, and applications, with applications accounting for only 5 percent of the revenues.

46. Agrium 2007, 44. The document lists numerous other sustainability efforts that are explicitly related to Agrium's manufacturing activities; I have not listed those here in this discussion (which focuses on distributors' retail operations).

47. CAPCA n.d.

48. Stoll 1998; Warner 2007, 245n22.

49. Van den Bosch 1978; Warner 2007, 245n22.

50. CAPCA n.d.

51. Warner (2007, 111) explains that farm advisers—university Cooperative Extension agents—are used "only for extraordinary knowledge needs," and that farm advice has become mostly privatized over the past few decades.

52. Ibid., 112–113. See also Epstein and Bassein 2003, 23.19.

53. Warner 2007, 113.

54. As indicated on the CAPCA (2009) Web site's recognition of its "diamond members."

55. Agrium 2008, 15.

56. For the trends in total amount of pesticides used in U.S. agriculture (measured in pounds of active ingredients), see U.S. EPA 2004, 28. For the trends in agricultural pesticides sales in the United States, see ibid., 23, although note that those figures have not been adjusted for inflation (33). Both of these metrics have their limitations. The trends in the amount of pesticides used do not account for variations in the toxicity of different pesticides (i.e., some exceptionally toxic pesticides are used in relatively low quantities), or the drift potential of different pesticide formulations or application methods; therefore, a downward trend in the overall pounds of pesticides used may not reflect a decrease in the toxic load into the environment. Pesticide sales data are similarly limited, since the price of a pesticide does not reflect its toxicity or drift potential.

57. Bosso 1987, 32–33.

58. See, for instance, Warner 2007; Hassenein 1999; Campbell 2001; Bell 2004.

59. Warner 2007, 97; Bell 2004; Warner 2007, 124.

60. For a history of organic farming in California, see Guthman 2004; Vos 2000. For the specific requirements of organic farming and food processing, see U.S. Department of Food and Agriculture 2011.

61. Cited in Fujii 2009.

62. For more information on the Spray Safe program, see Spray Safe (2011). For information on the Alliance for Food and Farming, see Alliance for Food and Farming (2011).

63. Ag Futures Alliance 2009.

64. For information, see National Coalition on Drift Minimization 2009.

65. National Agricultural Aviation Association 2009.

66. Whitney 2000.

67. On consolidation in off-farm sectors of the agrifood system, see Heffernan 2000; Howard 2009a, 2009b. For evidence of increasing farming input prices over the past ten years, see Lucier and Dettman 2008.

68. This paragraph describes the exploitative dynamics of capitalism. Marxist scholars have articulated these relationships in numerous ways; the "treadmill of production" (Gould, Pellow, and Schnaiberg 2004) is perhaps the most widely known of those in environmental sociology. Regarding the forces of global capitalism as the primary drivers of socioecological change, see Harvey 1996; Robbins 2004; Smith 1984.

69. To quote one farmer from a major national newspaper story on pesticide drift (Ritter 2005).

70. Warner 2007, 112.

71. Blaine et al. 2008; see table 10.

72. Carolan 2006.

73. Quandt et al. 1998; Perry and Bloom 1988.

74. The major grower and trade organizations tirelessly fight anything that can remotely be construed as critical of conventional agriculture—from alternative agriculture funds in the Farm Bill to Michelle Obama's organic garden at the White House (Beam 2009).

75. Bell 2004.

76. Ibid.; Burton 2004; Wilson 2001; Carolan 2006.

77. Carolan 2006.

78. Quandt et al. 1998; Martinez et al. 2004; Van Tassel et al. 1999; Larson 2000; Rao et al. 2004.

79. Shipp et al. 2007; Prado and Villarejo 1996, cited in Larson 2000, 19; Villarejo et al. 2000; Arcury, Quandt, and Russell 2002.

80. Larson 2000.

81. The county agriculture commissioner later fined the farmer over ten thousand dollars for this and other regulatory violations associated with the incident (Arroyo 2009a).

82. Per my public records request in August 2009, the California DPR gave me the regulatory violation information on all reported and confirmed pesticide illnesses due to pesticide drift in 2007. Of all pesticide drift illness cases in that year in which violations were found (though not including cases in which the only violation that investigators identified was the fact that drift occurred—i.e., violation of California Code of Regulations 6614, Food and Ag Code 12972, or Food and Ag Code 12973 with specific reference to allowing drift to occur), the majority of the cases were associated with violations of pesticide safety regulations. I elaborate on this discussion of regulatory violations in chapter 4.

83. DPR 2001a, 4.

84. Van Tassel et al. 1999.

85. California Code of Regulations 6600 stipulates in general terms that applicators must use "general standards of care" when applying pesticides; a violation of this code is often identified in pesticide drift illness investigations.

86. Cabrera and Leckie 2009; Austin et al. 2001; Arcury, Quandt, and Russell 2002.

87. For an interesting comparison and original data, see the survey of lawn care company employees in Robbins 2007.

88. Vaughan 1995.

89. Larson 2000.

90. DPR 2009f, 489.

91. The Kern County agriculture commissioner later fined the employer of the poisoned field crew a thousand dollars for this and other regulatory violations associated with the incident (Arroyo 2009b).

92. Khokha 2009.

93. Kroma and Flora 2003.

94. CropLife America 2009.

95. Interview with the author in 2004.

96. For critical accounts of the role of agriculture in the Central Valley's natural history, see Gottlieb 1988; Nash 2006; Reisner 1993; Arax and Wartzman 2005.

97. Ross 1998.

98. CropLife America 2009.

99. Patel, Torres, and Rosset 2005.

100. Bayer 2011.

101. CropLife America 2009.

102. DuPont 2009.

103. Statement at special legislative hearing on methyl iodide, Sacramento, CA, August 19, 2009.

104. Poppendieck 2000; Gottlieb and Joshi 2010; Winne 2008.

105. Farmer (and state representative) statement at MeI legislative hearing, Sacramento, CA, August 19, 2009.

106. Author interview with the president of a major growers organization in California.

107. Nozick 1974, 26.

108. Harvey 1996, 372.

109. National Agricultural Aviation Association 2009.

110. Senior vice president of Western Growers Association at special legislative hearing on methyl iodide, Sacramento, CA, August 19, 2009.

111. Young 1990, 28.

112. Harry Brighouse, personal conversation with author in 2008 in Madison, WI. For critiques of libertarians' treatments of environmental externalities, see Wenz 1988; Harvey 1996; Kymlicka 2002; Brighouse 2004. Although analysts who defend and apply libertarian philosophy typically recognize environmental externalities as worthy of government intervention, critics maintain that libertarians underestimate both the scope of negative externalities and the ability to easily address them in contexts of considerable inequality.

113. PAN 2009e.

114. See, for example, Beyond Pesticides 2009.

115. See Tyrone Hayes's damning testimony about the corporate links between agrichemical and pharmaceutical manufacturers in Land Stewardship Project and PAN 2010.

116. For the International Campaign for Justice in Bhopal, see International Campaign for Justice in Bhopal 2011. For the Dow Accountability Network, see Dow Accountability Network 2011. For the Agribusiness Action Initiatives (formerly known as the Agribusiness Accountability Initiative), see Agribusiness Action Initiatives 2011.

117. PAN 2009a, 2009f.

118. As stated in an email to other pesticide drift activists in 2005. The news article referenced here is Rodriguez 2005.

119. For an introduction to some of the major lawsuits brought against chemical manufacturers, see PAN 2009e.

Chapter 4

1. Love Canal is a neighborhood in New York State that became the site of considerable controversy after residents discovered in the 1970s that their homes and the local school had been built on a toxic waste dump. Through the work of residents and local reporters to investigate the waste, identify patterns of contamination and unreasonably high rates of illness, hold the responsible company accountable, and relocate the residents, the case gained national attention and has since been regarded as a defining moment in the U.S. environmental movement in general and antitoxics activism in particular (Gibbs 1982; Gottlieb 2005). The 1982 statement by Paigen was quoted in Brown et al. 2000, 20.

2. For a recent review of the different ways that social scientists theorize environmental governance, see Davidson and Frickel 2004. On regulatory science, see Jasanoff 1990; Irwin et al. 1997.

3. Bosso (1987) illustrates that this was first mandated through the Federal Insecticide Act of 1910, and that the Federal Insecticide, Fungicide, and Rodenticide Act (FIFRA) of 1947 elaborated this responsibility in only insignificant ways. See also Stoll 1998; Baker 1988.

4. Daniels 2007.

5. Stoll 1998, 115.

6. See DuPuis 2004; Tarr 1996.

7. Bosso 1987.

8. For elaborations of the history of agricultural policy, see McConnell 1966; Cochrane 1979; Strange 1988; Browne 1988, 1995; Winders 2009.

9. See Busch and Lacy 1983; Cochrane 1979; Henke 2008; Hightower 1973; Kleinman 2003; Kloppenburg 2005; Krimsky 2003; Nestle 2002; Strange 1988; Vallas and Kleinman 2008; Warner 2007.

10. Gottlieb 2005, 123; Whorton 1974.

11. Many scholars have addressed the profound (and complex) impact of *Silent Spring*. For a thoughtful discussion, see Gottlieb 2005.

12. Although FIFRA was first implemented in 1947, the amendments in 1972 greatly expanded the act's protections of human health and the environment. FIFRA has been further amended and expanded several times since.

13. Sunstein 2002.

14. Notably, California's pesticide regulatory program was moved out of the Department of Food and Agriculture and into the new California EPA's DPR in 1991, though DPR was primarily staffed by Department of Food and Agriculture employees. For the history of pesticide regulation in California, see DPR 2001b.

Forty-three U.S. states administer their pesticide regulatory programs through their departments of agriculture, and two others (South Carolina and Indiana) administer theirs through state universities. The only six states that house their pesticide regulatory programs within an institutional context of environmental protection are Alaska, California, Connecticut, New Jersey, New York, and Rhode Island.

15. Citing Jasanoff's account (2003) of the history of risk assessment within the regulatory arena, Whiteside (2006, 69) notes that a U.S. Supreme Court case from 1980 "made quantitative risk assessment essentially obligatory for all U.S. agencies involved in health regulation."

16. That "the label is the law" is one of FIFRA's key provisions.

17. DPR 2001b, 51.

18. Starr 1985; Pincetl 1999.

19. U.S. EPA 2009d.

20. DPR 2001b, 32.

21. Taylor 2004.

22. U.S. EPA 2008b, 2, 3.

23. For DPR staff and budget histories, see DPR 2009g. My estimate of "over 300" CAC staff is based on my estimate of 2,000 hours for a full-time equivalent employee and DPR's report that 662,090 staff hours were spent on all county pesticide activities in FY 2006–2007, as reported in the California Statewide Pesticide Regulatory Activities Summary of 2006–2007 (DPR 2007b).

24. Notable health-protective policies include the federal Food Quality Protection Act of 1996 (U.S. EPA 2010a) and California's Toxic Air Contaminant Act of 1984 (DPR 2009j).

25. U.S. EPA 2010a. Note, however, that the EPA still assumes that inhalation risk is zero except for "volatile" pesticides.

26. U.S. EPA 2009e.

27. PAN 2009c.

28. The present restricted use products process was instituted in FIFRA's revision in 1972. California's restricted materials process enforces this mandate and fulfills the environmental impact assessment component of the California Environmental Quality Act of 1976. An early version of California's relatively elaborate restricted materials program actually began in 1949 as an effort to protect crops from herbicide drift from neighboring fields (DPR 2001b, 7).

29. California Code of Regulations 2009.

30. See DPR 2008a, 2009i.

31. This program was suspended in 2002 due to budget cuts and then reinitiated in 2007.

32. For the critiques, see Welch et al. 2006. For a review of the EPA and the FQPA, see McGarity 2001.

33. U.S. EPA 2009g.

34. DPR 2008c. This ranking excludes the sulfur and horticultural oils because of their relatively minimal human toxicity.

35. The U.S. EPA's Web page (2009c) for methyl iodide includes sections titled "Stringent Protection Measures " and "One of the Most Thorough Risk Assessment Processes Ever Completed by the Agency." In contrast, see the letter of critique from scientists as well as environmental advocates' arguments on PAN's Web site (2011) on the pesticide.

36. Feldman et al. 2009. These regulations are still being negotiated as of 2011.

37. Atrazine was the top pesticide used in U.S. agriculture through 1999. In the most recent available national data, from 2001, it was surpassed by the herbicide glyphosate, or Roundup (U.S. EPA 2009g). For scientific indictments of its presence in human drinking water supplies, and impacts on human and nonhuman populations, see Hayes 2004; Land Stewardship Project and PAN 2010; Rohr and McCoy 2009; Union of Concerned Scientists 2009; Wu et al. 2009. Other particularly notable cases include captan and 2,4-D (both of which are widely used and have been linked to cancer in laboratory studies). The U.S. EPA (2009f) has pledged to reconsider the evidence against atrazine and other pesticides, but to date has not yet taken any action in that regard. The news media has also extensively criticized the U.S. EPA for its failure to restrict the use of atrazine (see, for example, Duhigg 2009).

38. The GAO's 2001 report makes this critique clearly: "In 1993, GAO reported that the existing sources of information on pesticide illnesses were limited in coverage, comprehensiveness, and quality. In that report, we concluded that without a valid system of monitoring pesticide illnesses, there was no way to identify problems that may occur with the different uses of pesticides or to determine whether practices intended to manage pesticide risks are effective in preventing hazardous exposure incidents. In our March 2000 report, we found that little had changed since 1993. While EPA uses four databases to provide some indication of the extent of pesticide illnesses, each of these databases has serious limitations. The lack of comprehensive nationwide data on pesticide illnesses remains largely unaddressed" (GAO 2001, 3).

39. Nash 2006.

40. DPR 2009j.

41. DPR 2009e.

42. DPR's PISP (2003, 6–7; 2004, 9; 2005, 10; 2006, 11–12; 2007, 10) annual summary reports from 2003 to 2007 indicate the number of agricultural pesticide illnesses in the PISP system and the subset of those for which violations had been determined. Regulatory violations that contributed to illness were identified in 1,188 (59 percent) of the 1,982 agricultural pesticide illnesses from 2003 to 2007. DPR staff members point out that the PISP database only includes data up to approximately fourteen months after the end of the calendar year in which pesticide illnesses occur. Thus, the database may not include violations that were determined much later in time, as can be the case in particularly contentious and/or lengthy illness investigations. The significance of this warning is unclear, as DPR representatives are not able to suggest the percentage of cases in which this might occur.

43. PISP 2007, 15; emphasis added.

44. PISP 2004, 10.

45. DPR staff members prepared the detailed data for 2007 for me per my public records request in August 2009. Due to time and budget constraints, DPR staff members were unfortunately only able to provide such data to me for 2007. In this data, the investigators were unable to identify any regulatory violations in 82 (47 percent) of the 175 PISP cases designated as definitely, probably, and possibly related to agricultural pesticide drift in 2007. (Specifically, the "violation" field for each of these 82 cases in the PISP database shows "none," "unknown," or "non-contributory," where the latter classification refers to cases in which a regulatory violation occurred but did not contribute to the illness.) In 22 (24 percent) of the 93 cases in which violations were identified, the only violation that investigators identified was the fact that drift took place.

46. In an additional 15 cases (16 percent of the cases in which a violation was identified), the only violations that the investigators determined were the fact that drift occurred (a violation of California Code of Regulations 6614, Food and Ag Code 12972, or Food and Ag Code 12973, with specific reference to allowing drift to occur) *and* that the applicator had not adhered to California Code of Regulations 6600 (which stipulates in broad terms that applicators must use general standards of care when applying pesticides).

47. DPR 2007a; emphasis added.

48. Bosso 1987.

49. Brickman, Jasanoff, and Ilgen 1985, 95.

50. NRC 1989, 83, 218.

51. Bosso 1987.

52. Raffensperger and Tickner 1999, 2.

53. For current examples of such critiques, see Dansereau 2009; Union of Concerned Scientists 2009; Fan 2003; Kegley, Katten, and Moses 2003; NRDC et al. 2003. See also the debate over methyl iodide, detailed in chapter 6. For historical accounts of such critiques, see Bosso 1987. For an accessible and illustrative critique of the risk assessment conducted on one soil fumigant, see Dansereau 2009. The major class of human health impacts that the U.S. EPA has systematically ignored is endocrine system disruption (Colborn, Dumanoski, and Myers 1997). The U.S. EPA did start requiring manufacturers to test pesticides for their effects on the human endocrine system in 2009, but even then it did so for only sixty-seven of the one thousand pesticides registered for use. Moreover, a leading endocrine researcher has critiqued the EPA's methodology as "outdated, insensitive, crude, and narrowly limited . . . and designed under the surveillance of corporate lawyers who had bottom lines to protect" (Theo Colborn, as cited in PAN 2009d, 3). For EPA scientists' own testimony of skipping crucial steps in the risk assessment process, see Welch et al. 2006.

54. Bosso 1987, 136, 199; Sass and Wu 2007.

55. Jasanoff 1992, 210.

56. Thornton 2000, 7. To ameliorate these problems, Thornton calls for restructuring chemical regulation around the "ecological paradigm." Compared to the pollution management basis of the risk paradigm, the ecological paradigm focuses on pollution prevention, is concerned with the total pollution burden to which people are exposed, accounts for the many factors that constrain the ability of science to accurately quantify chemical risk, and emphasizes the need for precaution-based, health-protective restrictions on chemicals. As I discuss later in this chapter, pesticide drift activists are fighting for regulatory reform in the spirit of the ecological paradigm.

57. PEER 2006c.

58. Sass and Wu 2007; PEER 2006a.

59. DPR 2006a.

60. DPR 2003.

61. DPR 2003, 34.

62. See Rankin Bohme, Zorabedian, and Egilman 2005; Sifry and Watzman 2004; Schlosser 2001; Spitzer 2005; Kovarik 2005; Snyder 2005; Sass 2005; Infante 2005; Rampton and Stauber 2002; Stauber and Rampton 1995. See also the work of Tyrone Hayes (2004) regarding the corporate control over his research findings that cast doubt on the safety of atrazine.

63. Bosso 1987, 132, 126.

64. The disproportionate concern for the impacts of pesticides on consumer health is evident in the U.S. EPA's early focus on environmentally persistent pesticides such as DDT, relatively radical (though ultimately unimplemented) legislation like the Delaney Clause of 1958 (which sought to ban all carcinogenic residues on food), and the more recent success with the FQPA.

65. DPR 2009h. For one egregious case that exemplifies both the revolving door and direct industry influence over policymaking, see Lardner and Warrick 2000.

66. Arguments between DPR and OEHHA over methyl bromide are a case in point here (Fan 2003).

67. Muzzling Those Pesky Scientists 2006.

68. Harris and Pear 2007; Unhealthy Influence 2007.

69. Welch et al. 2006; see also PEER 2006b.

70. Janofsky 2006.

71. Interview with the author. Note that I have omitted the names of the pesticides to protect the scientist's identity.

72. Stauber and Rampton 1995; Rampton and Stauber 2002.

73. Bosso 1987.

74. DPR 2009b. In fact, controversies over agriculture commissioners' unwillingness to issue fines eventually compelled DPR to develop and issue new enforcement response regulations in 2007 that codify the fines and other enforcement actions that agriculture commissioners must follow in cases of pesticide regulatory violations (California Code of Regulations 2009).

75. DPR 2009c.

76. Author's phone interview with Paul Gosselin in 2004. Gosselin estimated that "nine out of ten" permit conditions follow the statewide guidelines. Given his leadership role within DPR and demonstrated unwillingness to speak critically of the institution, I assume that this is an underestimate.

77. Note that this contradictory mandate is evident in CACs' mission statements. Kern County (2009) CAC's stated mission is representative: "To promote and protect the agricultural industry, environment, and general public."

78. I should note that in private, confidential conversations, some regulatory scientists deviate from this predominant story, admitting that regulatory programs do a poor job of evaluating numbers of pesticide exposures. I chose to emphasize the narrative of "accidents" here because it constitutes the public party line—the story that regulatory agency representatives consistently tell in their official capacity and public settings.

79. Quoted in Barbassa 2004.

80. Khokha 2009.

81. One of DPR's few recognitions of the limits of the PISP system is buried in a brief DPR (2009i, 2) handout on pesticide illnesses: "People injured off the job, or who fail to seek medical care after pesticide exposures, are unlikely to make it into the system. Reporting aspects of the program also tend to emphasize acute rather than chronic illnesses related to pesticide exposures. . . . Therefore, data should not be used to draw conclusions about the total number of pesticide illnesses."

82. DPR 2004.

83. PISP 2003, i.

84. There is a long history of officials dismissing critics of pesticides as "subjective" and "emotional." See, for example, the discussion of the reaction to *Silent Spring* in Bosso 1987, 120.

85. Used in this way, these narratives constitute a classic example of what sociologists of science call "boundary work": the practices through which scientists and other elites maintain a monopoly over authoritative knowledge in a particular field. In this case, "truth" is the purview of "disinterested" scientists, whose determinations differ distinctly from the "opinions" of other actors (Gieryn 1999; Allen 2004).

86. Garcia 2006; Khokha 2009; personal interviews with the author; Players Discuss Current System 2005.

87. Quoted in Barbassa 2005.

88. For a classic analysis of "normal accidents," see Perrow 1984.

89. A huge body of research in recent years has interrogated the forms and forces that together comprise the practices and ideology of neoliberalism, which refers to trends unfolding in the policy arena, institutional reforms, and individual mentalities of rule. A few key resources on the subject include Brenner and Theodore 2002; Castree 2008a, 2008b; Harvey 2005; Jessop 2002; Leitner, Peck, and Sheppard 2007; Peck and Tickell 2002; Peck 2004.

90. Leitner, Peck, and Sheppard 2007, 225.

91. NPR's *Morning Addition* with Mara Liasson on May 25, 2001.

92. Bird 1999, 139, cited in Lukes 2006, 10. See also Brighouse 2004, 84; Low and Gleeson 1998, 79.

93. DPR 2006b.

94. Fujii 2009. For a thoughtful analysis of some of the "neocommunitarian" programs in U.S. EPA's EJ initiative, see Holifield 2004.

95. Quoted in Khokha 2009.

96. For critiques of communitarian ideology in agrifood activism and elsewhere, see DuPuis and Goodman 2005; DuPuis, Goodman, and Harrison 2007; DuPuis, Goodman, and Harrison 2011; Cohen 2004; Slocum 2007.

97. Holifield 2004.

98. I elaborate on these critiques of communitarianism in chapter 5.

99. Resident-activist statement at legislative hearing on methyl iodide in Sacramento in August 2009, referring to the local agriculture commissioner's response to her concerns about methyl bromide applications made across the street from her family's house.

100. Pesticide drift activist, quoted in DeAnda 2006, 2.

101. For classic work on discursive frames in social movement theory, see Snow et al. 1986; Snow and Benford 1988, 1992; Benford and Snow 2000.

102. Personal interviews. Last quote here is from Littlefield 2009.

103. Statement at legislative hearing on methyl iodide, Sacramento, CA, August 2009.

104. Arguello et al. 2005, 4.

105. Quoted in CPR 2007, 10.

106. Statement at legislative hearing on methyl iodide, Sacramento, CA, August 2009.

107. For critical accounts of doctors' inabilities to accurately diagnose pesticide exposure, see Das et al. 2001; Pease et al. 1993; Reeves, Katten, and Guzmán 2002; Moses 1993; Moses et al. 1993.

108. Hsu 2003a, 2003b.

109. Quoted in Stapleton 2003, 21. For a series of instances that showcase regulatory failure to take pesticide illnesses seriously, see Reeves, Katten, and Guzmán 2002.

110. Kegley et al. 2009, 9.

111. See Raffensperger and Tickner 1999; Tickner 2003; Whiteside 2006.

112. To view scientist-activists' technical comment letters to the U.S. EPA on specific regulatory decisions, see each specific pesticide's file ("docket"), available at <http://www.epa.gov/pesticides/docket>.

113. Regarding EJ activists' adoption of the precautionary principle in recent years, see Brown 2007, 219.

114. This definition is from a letter from PAN's directors to its members, in which the authors framed PAN's current projects as united by the precautionary principle (Moore and Scholl-Buckwald n.d.).

115. Thornton 2000, 10.

116. Raffensperger and Tickner 1999. The Wingspread Statement, an outcome of the Wingspread Conference on the Precautionary Principle in 1998, summarizes the precautionary principle as follows: "Where an activity raises threats of harm to the environment or human health, precautionary measures should be taken even if some cause and effect relationships are not fully established scientifically" (Wingspread 1998).

117. Brown 2007, 205.

118. Whiteside 2006, 62.

119. Ibid., 57.

120. Raffensperger and Tickner 1999, 2.

121. Dansereau and Kegley 2007.

122. For strong, recent monographs on participatory science, see Allen 2003; Brown 2007; Corburn 2005; Sze 2007.

123. Sze 2007, 22.

124. O'Brien 2000.

125. This desire to reorient pesticide regulation away from industry protection and toward public health is what compels pesticide drift activists to regularly petition to have pesticide risk assessment shifted from DPR to OEHHA. Recall that industry pressure is what kept risk assessment in DPR rather than OEHHA, which has a much stronger culture of public health protection and conducts the risk assessments for all nonpesticide environmental pollutants (see the "Industry Influence" section in this chapter).

126. Dansereau 2009.

127. The FQPA has prompted the cancellation or partial ban of several organophosphates (Zalom, Toscano, and Byrne 2005; Van Steenwyk and Zalom 2005) and the development of many alternative pest management strategies in certain commodity sectors (especially in wine grapes, almonds, and pears; Warner 2007).

128. The Community Action Guide is available at CPR 2009a. Note that DPR (2008a) published a similar document ("Community Guide to Recognizing and Reporting Pesticide Problems") in 2008.

129. CPR 2010a.

130. The data collection began in 2007. See Northwest Coalition for Alternatives to Pesticides 2009.

131. To date, the Stockholm Convention has been signed by ninety-one countries plus the European Union.

132. See PAN 2010b; Roosevelt 2009; CRPE 2011b; DPR 2010e.

133. Goldman, Brimmer, and Ruiz 2009.

134. Recall that while the California EPA was being formed, industry pressured the state to locate pesticide risk assessment in DPR rather than OEHAA because it knew that DPR, whose staff members came largely from the state Department of Agriculture, would be more protective of industry interests. See the "Industry Influence" section earlier in this chapter.

135. See Kegley and Chatfield 2004, 4–5.

136. For PAN's campaign against the two recent industry appointees, see PAN 2010a.

137. See Dansereau 2009; Kegley, Katten, and Moses 2003, 4.

138. The Drift Catcher program was inspired by the success of the Bucket Brigades, a participatory air-monitoring program initially commissioned by Ed Masry and Erin Brockovich, and then further developed and implemented widely throughout the world by Communities for a Better Environment and the Global Community Monitor to document as well as combat air pollution around oil refineries (Bucket Brigade 2008; O'Rourke and Macey 2003; Ottinger 2010). For more on the Drift Catcher, see Harrison 2011; Kegley et al. 2009; Clarren 2008.

139. For an example of how Biodrift results are presented, see CPR 2007.

140. Young 1990, 41. See also Schlosberg 2007.

141. Fraser 2000, 113.

142. Young (1990, 58–59) defines cultural imperialism as cases in which "the dominant meanings of a society render the particular perspective of one's own group invisible at the same time as they stereotype one's group and make it out as Other."

143. Schlosberg (2007, 65–71) describes how participatory parity is a central tenet of the EJ movement (in terms of both its activist and academic wings).

144. The activist participants I interviewed were rather ambivalent about their participation in DPR EJ programs. For a critical and interesting discussion of a similar program administered by the U.S. EPA, see Holifield 2004.

145. See Cole and Foster 2001, chapter 5.

146. For a few excellent case studies of lay science, see research by Phil Brown and his colleagues (Brown and Mikkelsen 1990; Brown 2007), Barbara Allen (2003), and Jason Corborn (2005). The term "research silences" is from Brown 2007, 261–265. For my critical assessment of the Drift Catcher program, see Harrison 2011.

Chapter 5

1. For an engaging historical overview of the U.S. environmental movement, see Gottlieb 2005.

2. Gottlieb 2005; Pellow 2005; Tarr 1996; Taylor 2009.

3. Nash (2006) discusses residents' early concerns about pesticides and the ways that predominant understandings of the relationships between the body and the

environment prevent those private concerns from gaining traction or otherwise becoming public issues.

4. Pulido 1996b; Nash 2006; Cole and Foster 2001, 27.

5. Pulido 1996b, 71, 84.

6. Campbell 2001, 355.

7. All information in this paragraph is from the Community Alliance with Family Farmers' concise and remarkably telling self-history (2010), which is also summarized in Campbell 2001.

8. Brown and Getz 2008; Pulido 1996b; Majka and Majka 2000.

9. Allen et al. 2003, 68. For clarity, I removed a reference to Allen (1999) from the middle of this quote.

10. Organic Farming Research Foundation 2010. AMO stands for Asociación Mercado Orgánica (which translates in English to Organic Marketing Association).

11. The flagship characterization is per Guthman 2008c. Organic food and beverage sales in North America have been growing at approximately 20 percent annually (compared to 2–3 percent for general food and beverage sales) and generated twenty billion dollars in sales in 2007 (Organic Trade Association 2007).

12. Guthman 2004; Vos 2000; Belasco 1989.

13. Vos 2000, 246.

14. Guthman 2004; Howard 2009a, 2009b.

15. The Alar scandal in the 1980s, the FQPA of 1996, and the public's response to the National Organic Program Proposed Rule in 1997 are three noteworthy cases in point.

16. It is worth noting that organic is an imperfect system for less toxic agriculture, as there are many organically approved inputs that are toxic to humans. Sulfur is the classic example: it is a skin and eye irritant, can by highly prone to drift, and is extensively used. The toxicity is low, however, relative to the pesticides used in nonorganic production. Because there are many other pesticides that deserve serious interrogation for their impacts on public health, I do not problematize sulfur in this book.

17. The "eat food, mostly plants, and not too much" phrase is Pollan's slogan (2008).

18. Sabina 2010.

19. See DuPuis 2007; Guthman 2007a, 2007b; Bobrow-Strain 2007; Donohue 2009.

20. See Alkon and Agyeman 2011; Alkon and Norgaard 2009; Allen et al. 2003; Friedmann 2007; Gottlieb 2001; Gottlieb and Fisher 1996a, 1996b; Gottlieb and Joshi 2010; Poppendieck 1998; Winne 2008.

21. A roughly parallel movement internationally is the food sovereignty movement, whose participants call for food system reform that both protects small farmers and ensures affordable food for consumers. The food sovereignty movement

was spearheaded in the 1990s by Via Campesina, a global farmers network, out of concern about the ways that structural adjustment policies and other neoliberal institutions increase the power of capital over the lives of rural people as well as undermine human rights. Via Campesina calls for democratically determined food policy that both eradicates hunger and ensures (rather than undermines) sustainable domestic food production (providing farmers with access to seeds, water, markets, and credit; setting price floors; and protecting domestic producers from underpriced imports). Via Campesina (2011) argues that the food sovereignty movement "represents an alternative to neoliberal policies," and observers have hailed the food sovereignty movement as "a mass re-politicization of food politics" (Patel 2007, 91). I leave food sovereignty in the endnote here, since it largely unfolds beyond the United States. It is also worth mentioning that the food sovereignty movement has paid little attention to the environmental conditions of agricultural production, frequently romanticizes local communities, and typically idealizes family farmers.

22. Scholars note that projects are rarely able to achieve all these goals, and that projects are often unable to profitably compete with established businesses. For a thoughtful consideration of such difficulties, see Gottlieb 2001.

23. Ibid., 250–252.

24. Allen et al. 2003.

25. According to the latest USDA data, the 218,838 acres of certified organic harvested cropland accounts for 2.8 percent of all 7,633,173 acres of harvested cropland in California (per the USDA Organic Survey of 2008 and the Census of Agriculture in 2007).

26. Allen 2004.

27. Campbell 2001, 356.

28. Allen et al. 2003. See also Allen 2004; Getz, Brown, and Shreck 2008; Guthman 2004; Shreck, Getz, and Feenstra 2006.

29. Allen 2004, 80.

30. Ibid.; Guthman 2004; Naples 1994.

31. Williams 1973; Mitchell 1996.

32. Getz, Brown, and Shreck 2008.

33. Allen 2004.

34. Getz, Brown, and Shreck 2008, 490, 499.

35. Shreck, Getz, and Feenstra 2006.

36. The California Agricultural Labor Relations Act secures these rights for farmworkers in the state. Farmworkers do not have this right elsewhere in the United States, as they are explicitly exempted from the National Labor Relations Act.

37. Getz, Brown, and Shreck 2008.

38. Although Guthman (2004) along with Ron Strochlic and his colleagues (2008) found slightly higher wages on organic farms than on conventional ones, the latter study discovered that organic farmers offer fewer nonwage benefits (ibid., 25).

39. This change was reflected in congressional deliberations over the FQPA of 1996. Although a significant public health achievement that made pesticide risk assessments more stringent, the FQPA's passage was ultimately secured because of policymakers' concerns about the health effects that pesticide residues pose for consumers, rather than for farmworkers or others living in agricultural regions.

40. Brown and Getz 2008.

41. See the final paragraph of the March 2009 article in *Gourmet* magazine that attempts to instruct the reader on "Buying Slave-Free Fruits" (Estabrook 2009). One reader's published response is representative: "It's simple. If you are, like me, angry about human rights violation don't buy slave-labor tomatoes." The Fair Food Project in California implies this same suggestion (California Institute for Rural Studies 2010).

42. Farmworkers in the United States have long been exempt from many labor laws, including the right to overtime pay and the protection from employer retaliation when engaging in collective bargaining and strikes.

43. Allen and Sachs 1993.

44. As Allen (2004, 126) argues, this reliance on individualistic action is popular because its implications are tidy and inoffensive: "Perhaps one reason that ideologies of individualism are popular is that if social problems are treated as individual rather than social, everyone else can be absolved of complicity in contributing to or not helping to solve social problems."

45. What remains to be determined is whether market-based advocacy actually reinforces the hegemony of libertarianism—either ideologically or institutionally. By accepting rather than directly confronting the libertarian prescriptions, does market-based agrifood advocacy consequently strengthen the libertarian theory of justice and/or materially limit the politics of the possible? While it is clear that recent trends in pesticide advocacy comply with libertarian theories of justice, the questions of whether they in turn reinforce that theory (i.e., make it more prevalent in society) and/or limit the politics of the possible (i.e., make stronger regulations more difficult to achieve) are empirical ones that have not necessarily been answered. Scholarship that has recently tried to identify the impacts of trends in agrifood advocacy explicitly in terms of political theories of justice includes DuPuis and Goodman 2005; DuPuis, Goodman, and Harrison 2007; DuPuis, Harrison, and Goodman 2011. Other scholarship that critically interrogates libertarian ideals in agrifood advocacy, although without an explicit consideration of libertarianism as a theory of justice, includes Allen 1999; Allen et al. 2003; Allen and Kovach 2000; Allen and Guthman 2006; Brown and Getz 2008; Guthman 2008c, 2008d; Harrison 2008; Morris 2008; Pudup 2008. Counterarguments to such research include Kloppenburg and Hassanein 2006; Rudy 2006.

46. Szasz 2008.

47. See Pelletier et al. 1999.

48. See Hinrichs 2003. Some community food security projects are notable exceptions to this trend (see Gottlieb 2001).

49. Again, the community food security movement is a notable exception to this problem (ibid.).

50. See DuPuis, Harrison, and Goodman 2011; McWilliams 1973; Mitchell 2003; Pincetl 1999; Harrison 2006; Davis 1991. Such arguments are consistent with the findings of scholars who have studied localization politics in other sectors and places (see Cohen 2004).

51. DuPuis 2004, 2007; Bobrow-Strain 2007.

52. Guthman 2008a, 2008b; Slocum 2007.

53. For literature that interrogates the politics of the local food movement, see Allen 2004; Born and Purcell 2006; DuPuis and Goodman 2005; DuPuis, Goodman, and Harrison 2007; DuPuis, Harrison, and Goodman 2011; Hinrichs 2003.

54. Harvey 1996, 385, 384.

55. Pollan 2008, 160–161.

56. Young 1990, 12.

57. Gottlieb and Fisher (1996a, 1996b) and Alkon and Norgaard (2009) contend that the community food security movement and the broader "food justice" movement can serve as a theoretical and political bridge between sustainable agriculture and EJ.

58. Chatfield characterizes his current work as strongly influenced by his efforts with the UFW in the "tumultuous summers of 1973 and 1974." Chatfield also worked during the 1970s for the American Friends Service Committee's rural outreach education program, a project to educate people outside the Bay Area about the committee's social justice issues. He also helped start the American Friends Service Committee's Rural Education and Action Project, which critically examined the relationships between industrial agriculture and rural poverty.

59. It was formerly known as the California Sustainable Agriculture Working Group, although it folded recently.

60. CPR 2010b.

61. As discussed earlier in this chapter; see Getz, Brown, and Shreck 2008.

62. This tension was exemplified when one farmer who had previously agreed to allow a Drift Catcher project to be conducted in his fields later declined to participate, reportedly because of pressure from neighboring farmers.

63. Allen et al. 2003, 65.

64. Other researchers have examined activists' rejections of "polite" behavioral norms. See Bullard 1993; Capek 1993; Gibbs 1982; Brown and Mikkelson 1990.

65. DPR 2010b; London, Sze, and Lievanos 2008.

66. Cole and Foster 2001, 131. Giovanna Di Chiro (1998) argues that coalitions are a means through which EJ activists demonstrate that the "environment" is a part of all issues—as essential, not marginal.

67. CRPE 2011a.

68. For an excellent interrogation of activist struggles to decide whether to frame pollution conflicts as environmental injustices or more broadly, see Kurtz 2003.

Other scholarship on the EJ frame includes Capek 1993; Benford 2005; Taylor 2000. Several scholars worry that EJ loses its radical edge as activists expand the meaning of EJ to include a broad range of social concerns (Benford 2005; Pellow and Brulle 2005; Getches and Pellow 2002). Schlosberg disputes this notion, contending that the EJ movement's "expansive identity" is one of its key strengths and accomplishments. "Rather than dismiss the use of a broad environmental justice frame because it applies to so many issues, we should examine what it is about environmental justice that resonates with so many communities on so many issues. We should attempt to broaden our understanding of the frame and discourse" (Schlosberg 2007, 176).

69. The southern end of the Central Valley has some of the worst air quality in the nation. It has fallen into the status of "severe noncompliance" of Clean Air Act standards, and contains four of the nation's top five most ozone-polluted cities and four of the country's top ten metropolitan areas most polluted by year-round particle pollution (annual PM2.5) (American Lung Association 2010).

70. CRPE 2011b.

71. This pursuit of multiple alliances is evident in a recent effort in which groups endorsing the development of a "pesticide protection zone" in Tulare County included an eclectic collection of local, regional, and national air pollution organizations, EJ organizations, community groups, and economic justice groups (Aguirre 2006).

72. I recognize that some EJ groups and scholars promote a rather communitarian notion of justice. For example, Schlosberg (2007, 53) points out that Devon Peña often does this (e.g., Peña 2005). Although made in the spirit of undermining the status quo, such unreflexive localism falls into the same trap as more socially conservative communitarian projects.

Chapter 6

1. U.S. EPA 2008a; 2010b, 8.2.

2. Scholars Paul Mohai, David Pellow, and Timmons Roberts (2009) have called for increased critical evaluation of state agencies' EJ institutionalization efforts.

3. Clinton 1994.

4. For an overview of state EJ programs, see Bonorris 2007. The California EPA's EJ efforts are a relatively elaborate example. See California EPA 2010b. For a summary of the legislation that established the EJ framework in California, see California EPA 2010a. For a review of some of that work, see London, Sze, and Lievanos 2008; Sze et al. 2009; Shilling, London, and Lievanos 2009.

5. For a thoughtful discussion of some of the positive contributions (and limitations) of these institutionalization efforts, see Cole and Foster 2001, 161–164.

6. Gottlieb 2001, xiii; see also Cole and Foster 2001.

7. See Bullard 2006; GAO 2005; U.S. EPA 2006. For some initial academic reactions to Executive Order 12898 that suggest the need for skepticism, see Benford

2005; Cable and Shriver 1995; Cole and Foster 2001; Gottlieb 2001; Harvey 1996; Sandweiss 1998.

8. Raju 2005.

9. Holifield 2004, 285.

10. Ibid., 296; Holifield, Porter, and Walker 2009, 598.

11. Cole and Foster 2001, 161.

12. Brown 2007, 220–221; Faber 2008; Morello-Frosch et al. 2006; Sze 2007. Andrew Szasz (1994) also defines pollution prevention as central to EJ.

13. Brown 2007, 205–207.

14. Similarly, Brown (2007, 207) argues that EJ constitutes a "new impetus to further development of the [precautionary] principle."

15. Gottlieb 1995.

16. Whiteside 2006, 63.

17. Per my own observations and those of several activist participants in the process.

18. Statement by Barry Underhill at a legislative hearing on methyl iodide, Sacramento, CA, August 2009.

19. Brown 2007, 212.

20. See ibid., 210–211; O'Brien 2000, 165–167; Whiteside 2006, 70–87.

21. European Union 2011.

22. Quote was made by John D. Graham (2002, quoted in Whiteside 2006, 63), a former administrator in the U.S. Office of Management and Budget. Regarding the long-standing roots of precautionary standards in U.S. environmental and public health legislation, see also Brown 2007, 208–209; Jasanoff 2003; Langston 2010; O'Brien 2000, 147–167; Raffensperger and Tickner 1999, 4–7; Whiteside 2006.

23. As O'Brien (2000, 80–81) observes, FIFRA "precludes consideration of alternatives (e.g., that more benign alternatives exist) when registering a pesticide for sale and use in the United States (7 U.S.C.A, Section 136a(c)(5)): 'The [EPA] Administrator shall not make any lack of essentiality a criterion for denying registration of any pesticide.'" O'Brien then points to several legal arguments that could justify revising this FIFRA language and institutionalizing alternatives assessment more broadly in U.S. environmental policy.

24. The California Environmental Quality Act's alternatives assessment mandate is codified in the California Code of Regulations, title 14, section 15126.6.

25. This is codified in the California Food and Agriculture Code, section 12825.

26. The FQPA requires that the U.S. EPA (2009e) revise its risk assessment procedures for all pesticides to better "consider aggregate risks from the same pesticide used in agricultural, commercial and/or residential settings, cumulative risks from exposure to pesticides with common mechanisms of toxicity, and the unique risks posed to infants and children due to their potentially increased sensitivity to pesticides." The FQPA's relatively narrow interpretation of what cumulative risk assessment means is one that was relatively easy for EPA scientists to operationalize

at the time that the law was passed. I discuss cumulative risk assessment again later in this chapter.

27. California Health and Safety Code, section 39650(b).

28. Occupational Safety and Health Administration 1998.

29. Weinberg, Bunin, and Das 2009, 56.

30. Johansen 2010; Serafini 2009.

31. PAN 2011.

32. For a summary of the scientific review committee's concerns about the DPR risk assessment as of February 2010, see Froines 2010.

33. The first two quotes come from the legislative hearing on methyl iodide, Sacramento, CA, June 2010. The second two quotes are from a scientific review committee public hearing on methyl iodide, September 2009 (DPR 2009f, 318–319).

34. Froines 2010, 5.

35. California's "mill tax" is one example, although a federal tax should be higher than the current mill tax (as of early 2010, it was only a 2.1 percent), tiered according to toxicity along with the risk of air and water contamination, and more fully dedicated than the California mill tax presently is to funding alternative pest management research and outreach programs.

36. PAN (2010c, 26–27) makes this argument in its comment letter to DPR on its proposed methyl iodide mitigation measures. Another interpretation of cumulative risk assessment might call for designing regulations to limit the overall regional loads of all pesticides that share a common mode of toxicity, which is how the EPA has interpreted the FQPA's cumulative risk assessment requirement.

37. U.S. EPA 2009h.

38. PAN (2010c, 25) made this argument in its comment letter to DPR on its proposed methyl iodide mitigation measures.

39. Allen (2003, 155–158). The literature on public participation in science is vast. For one recent review and typology of different models, see Rowe and Frewer 2005.

References

Ag Futures Alliance. 2009. Alliance for Food and Farming. <http://www .foodandfarming.info> (accessed October 26, 2009).

Agribusiness Action Initiatives. 2011. <http://www.agribusinessaction.org> (accessed January 16, 2011).

Agrium. 2007. 2007 Sustainability Report: A Growing Impact. <http://www .agrium.com/sustainability_report/agrium_sr07/inside/about_report.html> (accessed January 14, 2011).

Agrium. 2008. 2008 Annual Report: Fundamentals of Growth. <http://www .agrium.com/investors/2008_annual_report.jsp> (accessed January 14, 2011).

Aguirre, Gustavo. 2006. Email Update of Safe Air for Everyone, Tulare County Pesticide Protection Zone Campaign. November 13.

Aguirre International. 2005. The California Farm Labor Force: Overview and Trends from the National Agricultural Workers Survey. June.

Agyeman, Julian, Robert D. Bullard, and Bob Evans, eds. 2003. *Just Sustainabilities: Development in an Unequal World*. Cambridge, MA: MIT Press.

Akesson, Norman B., and Wesley E. Yates. 1964. Problems Relating to Application of Agricultural Chemicals and Resulting Drift Residues. *Annual Review of Entomology* 9: 285–318.

Alkon, Alison Hope, and Julian Agyeman, eds. 2011. *Cultivating Food Justice*. Cambridge, MA: MIT Press.

Alkon, Alison Hope, and Kari Marie Norgaard. 2009. Breaking the Food Chains: An Investigation of Food Justice Activism. *Sociological Inquiry* 79 (3): 289–305.

Allen, Barbara. 2003. *Uneasy Alchemy: Citizens and Experts in Louisiana's Chemical Corridor Disputes*. Cambridge, MA: MIT Press.

Allen, Patricia. 1999. Reweaving the Food Security Safety Net: Mediating Entitlement and Entrepreneurship. *Agriculture and Human Values* 16 (2): 117–129.

Allen, Patricia, ed. 1993. *Food for the Future: Conditions and Contradictions of Sustainability*. New York: Wiley.

Allen, Patricia. 2004. *Together at the Table: Sustainability and Sustenance in the American Agrifood System*. University Park: Pennsylvania State University Press.

Allen, Patricia, Margaret FitzSimmons, Michael Goodman, and Keith Warner. 2003. Shifting Plates in the Agrifood Landscape: The Tectonics of Alternative Agrifood Initiatives in California. *Journal of Rural Studies* 19 (1): 61–75.

Allen, Patricia, and Julie Guthman. 2006. From "Old School" to "Farm-to-School": Neoliberalization from the Ground Up. *Agriculture and Human Values* 23 (4): 401–415.

Allen, Patricia, and Marty Kovach. 2000. The Capitalist Composition of Organic: The Potential of Markets in Fulfilling the Promise of Organic Agriculture. *Agriculture and Human Values* 17 (3): 221–232.

Allen, Patricia, and Carolyn Sachs. 1993. Sustainable Agriculture in the United States: Engagements, Silences, and Possibilities for Transformation. In *Food for the Future: Conditions and Contradictions of Sustainability*, ed. Patricia Allen, 139–167. New York: John Wiley and Sons.

Alliance for Food and Farming. 2011. <http://www.foodandfarming.info/about .asp> (accessed January 4, 2011).

Als-Nielsen, Bodil, W. Chen, C. Gluud, and Lise L. Kjaergard. 2003. Association of Funding and Conclusions in Randomized Drug Trials: A Reflection of Treatment Effect or Adverse Events. *Journal of the American Medical Association* 290: 921–928.

Altieri, Miguel. 1987. *Agroecology: The Scientific Basis of Alternative Agriculture.* Boulder, CO: Westview Press.

Alvarez, Fred. 2000a. Parents Call for Ban on Pesticide Use Near Schools. *Los Angeles Times*, November 10, B1.

Alvarez, Fred. 2000b. State Tests Find Pesticide on Campus. *Los Angeles Times*, November 15, B1.

American Lung Association. 2010. Most Polluted Cities. American Lung Association State of the Air 2010. <http://www.stateoftheair.org/2010/city-rankings /most-polluted-cities.html> (accessed January 14, 2011).

Andreas, Peter. 2000. *Border Games.* Ithaca, NY: Cornell University Press.

Andreas, Peter. 2001. The Transformation of Migrant Smuggling across the U.S.-Mexican Border. In *Global Human Smuggling: Comparative Perspectives*, ed. David Kyle and Rey Koslowski, 107–125. Baltimore: Johns Hopkins Press.

Arax, Mark, and Rick Wartzman. 2005. *The King of California: J.G. Boswell and the Making of a Secret American Empire.* New York: PublicAffairs Books.

Arcury, Thomas A., Sara A. Quandt, and Gregory B. Russell. 2002. Pesticide Safety among Farmworkers: Perceived Risk and Perceived Control as Factors Reflecting Environmental Justice. *Environmental Health Perspectives* 110 (supplement 2): 233–239.

Arguello, Martha Dina, et al. 2005. Comments to US EPA on the Fumigant Cluster Assessment. October 12. <http://www.panna.org/fumigants/review#Phase 3Comments>.

Arroyo, Ruben J. 2009a. Notice of Proposed Action, Grounds Thereof, and of Opportunity to Be Heard. Kern County Department of Agriculture and Measurement Standards. File number 009-ACP-KER-09/10. October 6.

Arroyo, Ruben J. 2009b. Notice of Proposed Action, Grounds Thereof, and of Opportunity to Be Heard. Kern County Department of Agriculture and Measurement Standards. File number 010-ACP-KER-09/10. October 6.

Aston, Linda S., and James N. Seiber. 1997. Fate of Summertime Airborne Organophosphate Pesticide Residues in the Sierra Nevada Mountains. *Journal of Environmental Quality* 26 (6): 1483–1492.

Austin, Colin K., Thomas A. Arcury, Sara A. Quandt, John S. Preisser, Luis F. Cabrera, and Rosa M. Saavedra. 2001. Training Farmworkers about Pesticide Safety: Issues of Control. *Journal of Health Care for the Poor and Underserved* 12 (2): 236–249.

Baker, Brian. 1988. Pest Control in the Public Interest: Crop Protection in California. *Journal of Environmental Law and Policy* 8 (1): 31–72.

Baker, Lynton W., Donald L. Fitzell, James N. Seiber, Thomas R. Parker, Takayuki Shibamoto, Michael W. Poore, Karl E. Longley, Ruth P. Tomlin, Ralph Propper, and David W. Duncan. 1996. Ambient Air Concentrations of Pesticides in California. *Environmental Science and Technology* 30 (4): 1365–1368.

Barbassa, Juliana. 2004. Little Help for Workers Exposed to Pesticides: 19 Recently Sickened in Drift Incident. *San Diego Union-Tribune*, May 22.

Barbassa, Juliana. 2005. Activists Say Large Settlements, Fines, Send Message to Pesticide Applicators. *San Diego Union-Tribune*, November 23.

Bartley, Tim. 2007. Institutional Emergence in an Era of Globalization: The Rise of Transnational Private Regulation of Labor and Environmental Conditions. *American Journal of Sociology* 113 (2): 297–351.

Bayer. 2011. Bayer CropScience. <http://www.bayercropscience.com/bcsweb/crop protection.nsf/id/en_the_second_green_revolution> (accessed January 14, 2011).

Beam, Christopher. 2009. Organic Panic: Michelle Obama's Garden and Its Discontents. *Slate.com*, June 4. <http://www.slate.com/id/2219772> (accessed December 26, 2009).

Becker, Jonathan, William Chism, Monisha Kaul, David Donaldson, and Tim Kiely. 2005. Overview of the Use and Usage of Soil Fumigants, U.S. Environmental Protection Agency, June 15.

Belasco, Warren. 1989. *Appetite for Change: How the Counterculture Took on the Food Industry*. New York: Pantheon Books.

Bell, Michael. 2004. *Farming for Us All: Practical Agriculture and the Cultivation of Sustainability*. University Park: Pennsylvania State University Press.

Benbrook, Charles. 1996. *Pest Management at the Crossroads*. Yonkers, NY: Consumers Union.

Benbrook, Charles. 2009. Impacts of Genetically Engineered Crops on Pesticide Use: The First Thirteen Years. Critical Issues Report. The Organic Center. <http://www.organic-center.org/science.tocreports.html> (accessed January 16, 2011).

Benford, Robert D. 2005. The Half-life of the Environmental Justice Frame: Innovation, Diffusion, and Stagnation. In *Power, Justice, and the Environment: A Critical Appraisal of the Environmental Justice Movement*, ed. David N. Pellow and Robert J. Brulle, 37–53. Cambridge, MA: MIT Press.

Benford, Robert D., and David A. Snow. 2000. Framing Processes and Social Movements: An Overview and Assessment. *Annual Review of Sociology* 26: 611–639.

Bergman, Robert G., et al. 2007. Letter to US EPA Administrator Stephen Johnson. September 24. <http://www.pesticidereform.org/article.php?id=342> (accessed December 21, 2009).

Beyond Pesticides. 2009. Genetic Engineering. <http://www.beyondpesticides.org /gmos/index.htm> (accessed December 26, 2009).

Bird, Colin. 1999. *The Myth of Liberal Individualism.* Cambridge: Cambridge University Press.

Blaine, Thomas W., Franklin R. Hall, Roger A. Downer, and Timothy Ebert. 2008. An Assessment of Agricultural Producers' Attitudes and Practices concerning Pesticide Spray Drift: Implications for Extension Education. *Journal of Extension* 46 (4). <http://www.joe.org/joe/2008august/a8.php> (accessed January 14, 2011).

Bobrow-Strain, Aaron. 2007. Kills a Body Twelve Ways: Bread Fear and the Politics of "What to Eat?" *Gastronomica* 7 (2): 45–52.

Bonanno, Alessandro, Lawrence Busch, William H. Friedland, Lourdes Gouveia, and Enzo Mingione, eds. 1994. *From Columbus to Conagra.* Lawrence: University Press of Kansas.

Bonorris, Steven, ed. 2007. *Environmental Justice for All: A Fifty State Survey of Legislation, Policies, and Cases.* 3rd ed. San Francisco: Public Law Research Institute, University of California Hastings College of the Law.

Born, Branden, and Mark Purcell. 2006. Avoiding the Local Trap: Scale and Food Systems in Planning Research. *Journal of Planning Education and Research* 26 (2): 195–207.

Bosso, Christopher. 1987. *Pesticides and Politics: The Life Cycle of a Public Issue.* Pittsburgh: University of Pittsburgh Press.

Boyd, C. A., M. H. Weiler, and W. P. Porter. 1990. Behavioral and Neurochemical Changes Associated with Chronic Exposure to Low-level Concentrations of Pesticide Mixtures. *Journal of Toxicology and Environmental Health* 30 (3):209–221.

Brenner, Neil, and Nik Theodore. 2002. Cities and the Geographies of "Actually Existing Neoliberalism." *Antipode* 34 (3): 349–379.

Brickman, Ronald, Sheila Jasanoff, and Thomas Ilgen. 1985. *Controlling Chemicals: The Politics of Regulation in Europe and the United States.* Ithaca, NY: Cornell University Press.

Brighouse, Harry. 2004. *Justice.* Cambridge: Polity.

Brown, Phil. 2000. Popular Epidemiology and Toxic Waste Contamination: Lay and Professional Ways of Knowing. In *Illness and the Environment: A Reader in Contested Medicine,* ed. Steve Kroll-Smith, Phil Brown, and Valerie J. Gunter, 364–383. New York: New York University Press.

Brown, Phil. 2007. *Toxic Exposures: Contested Illnesses and the Environmental Health Movement.* New York: Columbia University Press.

Brown, Phil, Steve Kroll-Smith, and Valerie J. Gunter. 2000. Knowledge, Citizens, and Organizations: An Overview of Environments, Diseases, and Social Conflict. In *Illness and the Environment: A Reader in Contested Medicine*, ed. Steve Kroll-Smith, Phil Brown, and Valerie J. Gunter, 364–383. New York: New York University Press.

Brown, Phil, and Edwin Mikkelsen. 1990. *No Safe Place: Toxic Waste, Leukemia, and Community Action*. Berkeley: University of California Press.

Brown, Sandy, and Christy Getz. 2008. Privatizing Farm Worker Justice: Regulating Labor through Voluntary Certification and Labeling. *Geoforum* 39 (3):1184–1196.

Browne, William P., Jerry R. Skees, Louis E. Swanson, Paul B. Thompson, and Laurian J. Unnevehr. 2002. *Sacred Cows and Hot Potatoes: Agrarian Myths in Agricultural Policy*. Boulder, CO: Westview.

Browne, William P. 1988. *Private Interests, Public Policy, and American Agriculture*. Lawrence: University Press of Kansas.

Browne, William P. 1995. *Cultivating Congress: Constituents, Issues, and Interests in Agricultural Policymaking*. Lawrence: University Press of Kansas.

Bryant, Bunyan. 1995. *Environmental Justice: Issues, Policies, and Solutions*. Covelo, CA: Island Press.

Bryant, Bunyan, and Paul Mohai, eds. 1992. *Race and the Incidence of Environmental Hazards: A Time for Discourse*. Boulder, CO: Westview Press.

Bucket Brigade. 2008. Bucket Brigade Introduction. <http://www.bucketbrigade.net/article.php?list=type&type=9> (accessed June 2, 2008).

Bullard, Robert D. 1990. *Dumping in Dixie: Race, Class, and Environmental Quality*. Boulder, CO: Westview Press.

Bullard, Robert D., ed. 1993. *Confronting Environmental Racism: Voices from the Grassroots*. Boston: South End Press.

Bullard, Robert D., ed. 2005. *The Quest for Environmental Justice: Human Rights and the Politics of Pollution*. San Francisco: Sierra Club Books.

Bullard, Robert D. 2006. New IG Study Blasts EPA's Environmental Justice Record: Latest Study Follows a String of Reports Giving EPA a Failing Grade on EJ. Environmental Justice Resource Center. September 25. <http://www.ejrc.cau.edu/IGReportOnEJ.htm> (accessed December 21, 2009).

Burton, Rob. 2004. Seeing Through the "Good Farmer's" Eyes: Towards Developing an Understanding of the Social Symbolic Value of "Productivist" Behaviour. *Sociologia Ruralis* 44 (2): 195–215.

Busch, Lawrence, and William B. Lacy. 1983. *Science, Agriculture, and the Politics of Research*. Boulder, CO: Westview Press.

Busch, Lawrence, William B. Lacy, Jeffrey Burkhardt, and Laura R. Lacy. 1991. *Plants, Power, and Profit: Social, Economic, and Ethical Consequences of the New Biotechnologies*. Cambridge, MA: Blackwell.

Cable, Sherry, and Thomas Shriver. 1995. The Production and Extrapolation of Meaning in the Environmental Justice Movement. *Sociological Spectrum* 15: 419–442.

Cabrera, Nolan L., and James O. Leckie. 2009. Pesticide Risk Communication, Risk Perception, and Self-protective Behaviors among Farmworkers in California's Salinas Valley. *Hispanic Journal of Behavioral Sciences* 31 (2): 258–272.

California Code of Regulations. 2009. Division 6: Pesticides and Pest Control Operations. Chapter 1: Pesticide Regulatory Program. Subchapter 3: Agricultural Commissioner Penalties. Article 1: Guidelines. 6128 (Enforcement Response) and 6130 (Civil Penalty Actions by Commissioners). <http://www.cdpr.ca.gov/docs /legbills/calcode/010301.htm#a6128> (accessed July 9, 2009).

California EPA (Environmental Protection Agency). 2010a. Environmental Justice Legislation. <http://www.calepa.ca.gov/EnvJustice/Legislation> (accessed June 12, 2010).

California EPA. 2010b. Environmental Justice Program. <http://www.calepa .ca.gov/EnvJustice> (accessed June 12, 2010).

California Institute for Rural Studies and Rick Nahmias Photography. 2010. Fair Food Project. <http://www.fairfoodproject.org/main> (accessed January 2, 2010).

California: Prison, Air, Budget. 2010. *Rural Migration News* 16 (3). <http:// migration.ucdavis.edu> (accessed June 23, 2010).

California Secretary of State. 2009. California Secretary of State Lobbying Activity. <http://cal-access.sos.ca.gov/Lobbying> (accessed October 2, 2009).

Camacho, David, ed. 1998. *Environmental Injustices, Political Struggles: Race, Class, and the Environment.* Durham, NC: Duke University Press.

Campbell, David. 2001. Conviction Seeking Efficacy: Sustainable Agriculture and the Politics of Co-optation. *Agriculture and Human Values* 18 (4):353–363.

CAPCA (California Association of Pest Control Advisers). n.d. Position Paper: Conflict of Interest. California Association of Pest Control Advisers. <http://capca .com/positionpapers> (accessed December 26, 2009).

CAPCA. 2009. <http://capca.com/links> (accessed October 12, 2009).

Capek, Stella. 1993. The "Environmental Justice" Frame: A Conceptual Discussion and an Application. *Social Problems* 40 (1): 5–24.

Carolan, Michael. 2006. Do You See What I See? Epistemic Barriers to Sustainable Agriculture. *Rural Sociology* 71 (2): 232–260.

Carson, Rachel. 1962. *Silent Spring.* Boston: Houghton Mifflin.

Castorina, Rosemary, and Tracey J. Woodruff. 2003. Assessment of Potential Risk Levels Associated with US Environmental Protection Agency Reference Values. *Environmental Health Perspectives* 111 (10): 1318–1325.

Castree, Noel. 2008a. Neoliberalising Nature: The Logics of Deregulation and Reregulation. *Environment and Planning A* 40 (1): 131–152.

Castree, Noel. 2008b. Neoliberalising Nature: Processes, Effects, and Evaluations. *Environment and Planning A* 40 (1): 153–173.

Center for Media and Democracy. 2011. <http://www.prwatch.org> (accessed 4 January 2011).

Center for Responsive Politics. 2009. <http://www.opensecrets.org/influence/index .php> (accessed October 2, 2009).

Charizopoulos, Emmanouil, and Euphemia Papadopoulou-Mourkidou. 1999. Occurrence of Pesticides in Rain of the Axios River Basin, Greece. *Environmental Science and Technology* 33 (14): 2363–2368.

Chavez, Leo. 2001. *Covering Immigration: Popular Images and the Politics of the Nation.* Berkeley: University of California Press.

Clarren, Rebecca. 2008. Pesticide Drift: Immigrants in California's Central Valley Are Sick of Breathing Poisoned Air. *Orion*, July–August, 56–63.

Clinton, William. 1994. Executive Order 12898 of February 11, 1994: Federal Actions to Address Environmental Justice in Minority Populations and Low-Income Populations. *Federal Register* 59 (32), February 16.

Coalition for Chemical Safety. 2009. <http://www.coalitionforchemsafety.com /default.aspx> (accessed November 6, 2009).

Cochrane, Willard. 1979. *The Development of American Agriculture: A Historical Analysis.* Minneapolis: University of Minnesota Press.

Cohen, Lisbeth. 2004. *A Consumer's Republic: The Politics of Mass Consumption in Postwar America.* New York: Vintage Books.

Colborn, Theo, Dianne Dumanoski, and John P. Myers. 1997. *Our Stolen Future.* New York: Penguin.

Cole, Luke W., and Sheila R. Foster. 2001. *From the Ground Up: Environmental Racism and the Rise of the Environmental Justice Movement.* New York: New York University Press.

Coleman, Mathew. 2007. Immigration Geopolitics beyond the Mexico-U.S. Border. *Antipode* 39 (1): 54–76.

Community Alliance with Family Farmers. 2010. History. <http://www.caff.org /join/history.shtml> (accessed January 2, 2010).

Corburn, Jason. 2005. *Street Science: Community Knowledge and Environmental Health Justice.* Cambridge, MA: MIT Press.

Corley, Rob. 2000. Opinion Piece: Who Will Protect the Children at School? *Los Angeles Times*, November 19, B17.

Cornelius, Wayne A. 2001. Death at the Border: The Efficacy and "Unintended" Consequences of U.S. Immigration Control Policy. *Population and Development Review* 27 (4): 661–685.

Couto, Richard A. 1985. Failing Health and New Prescriptions: Community-Based Approaches to Environmental Risks. In *Current Health Policy Issues and Alternatives: An Applied Social Science Perspective*, ed. Carole E. Hill. Athens: University of Georgia Press.

Cowan, Tadlock. 2005. "California's San Joaquin Valley: A Region in Transition." Congressional Research Service Report for Congress. <http://fpc.state.gov /documents/organization/59030.pdf> (accessed January 16, 2011).

CPR (Californians for Pesticide Reform). 2007. Airborne Poisons: Pesticides in Our Air and in Our Bodies: Profiles of Lindsay Residents Who Participated in the Study. <http://pesticidereform.org/article.php?id=297&preview=1&cache=0> (accessed October 2, 2009).

CPR. 2009a. The Threat of Pesticides in Our Air: A Community Response Guide. <http://www.pesticidereform.org/article.php?list=type&type=22> (accessed December 30, 2009).

CPR. 2009b. Tulare County Pesticide Drift Survey Results. <http://www.pesticide reform.org/downloads/SurveyResults-Summary.pdf> (accessed July 20, 2009).

CPR. 2010a. Buffer Zone Update: Stanislaus Victory! CPR Update Newsletter. June 3.

CPR. 2010b. Healthy Harvest: From Field to Table. Californians for Pesticide Reform Annual Conference. March 20, 2010. Sacramento, CA.

Cronon, William. 1990. Modes of Prophecy and Production: Placing Nature in History. *Journal of American History* 76 (4): 1122–1131.

Cronon, William. 1992. A Place for Stories: Nature, History, and Narrative. *Journal of American History* 78 (4): 1347–1376.

Cronon, William. 1998. The Trouble with Wilderness, or, Getting Back to the Wrong Nature. In *The Great New Wilderness Debate*, ed. J. Baird Callicott and Michael P. Nelson, 471–499. Athens: University of Georgia Press.

CropLife. 2008. CropLife 100 List of Top 100 Agricultural Retailers in the United States. <http://www.croplife.com/retailers/croplife100/?year=2008> (accessed August 12, 2009).

CropLife America. 2009. <http://www.croplifeamerica.com> (accessed November 8, 2009).

CRPE (Center on Race, Poverty, and the Environment). 2011a. <http://www.crpe-ej .org/crpe> (accessed January 14, 2011).

CRPE. 2011b. Clean Air in the San Joaquin Valley: CRPE Protects the Valley's Most Vulnerable Residents by Making the Largest Polluters Abide by the Clean Air Act. <http://www.crpe-ej.org/crpe/index.php?option=com_content&view =article&id=59&Itemid=66> (accessed January 14, 2011).

Daniel, Cletus. 1981. *Bitter Harvest: A History of California Farm Workers, 1870–1941*. Ithaca, NY: Cornell University Press.

Daniels, Pete. 2007. *Toxic Drift: Pesticides and Health in the Post–World War II South*. Baton Rouge: Louisiana State University Press.

Dansereau, Carol. 2009. MITC in Our Air: Air Monitoring Results, Poisoning Cases, and the Need for Action to Protect Workers and Families. Farm Worker Pesticide Project. <http://www.fwpp.org> (accessed July 28, 2010).

Dansereau, Carol, and Susan Kegley. 2007. Letter submitted to editor of the *Seattle Post Intelligencer*.

Das, Rupali, Andrea Steege, Sherry Barron, John Beckman, and Robert Harrison. 2001. Pesticide-Related Illness among Migrant Farm Workers in the United States. *International Journal of Occupational and Environmental Health* 7 (4): 303–312.

DataMonitor. 2008. Fertilizers and Agricultural Chemicals: Global Industry Guide. Report Summary. <http://www.reportlinker.com/p099566/Fertilizers -Agricultural-Chemicals-Global-Industry-Guide.html> (accessed September 11, 2009).

Davidson, Carlos. 2004. Declining Downwind: Amphibian Population Declines in California and Historic Pesticide Use. *Ecological Applications* 14 (6): 1892–1902.

Davidson, Carlos, H. Bradley Shaffer, and Mark R. Jennings. 2002. Spatial Tests of the Pesticide Drift, Habitat Destruction, UV-B, and Climate Change Hypotheses for California Amphibian Declines. *Conservation Biology* 16 (6): 1588–1601.

Davidson, Debra J., and Scott Frickel. 2004. Understanding Environmental Governance: A Critical Review. *Organization and Environment* 17 (4): 471–492.

Davis, Mike. 1991. *City of Quartz*. London: Verso.

DeAnda, Teresa. 2006. Community Members Compensated for Pesticide Drift Poisoning: Three Years Later, Arvin Residents Win Settlement Against Applicator, Farmer. CPR Resource no. 16 (March). <http://www.pesticidereform.org/article .php?id=251> (accessed January 14, 2011).

De Guzman, Doris. 2009. Fertile Ground. *ICIS Chemical Business*, July 13–19, 276.

Delaney, Arthur. 2010. Lobbying's New Frontier: "Not Lobbying." *Huffington Post*, January 6. <http://www.huffingtonpost.com/2010/01/06/lobbyings-new -frontier-no_n_411639.html> (accessed January 14, 2011).

Department of Homeland Security. 2009. Snapshot: A Summary of CBP Facts and Figures. U.S. Customs and Border Patrol. <http://www.cbp.gov/xp/cgov/about /accomplish> (accessed May 16, 2010).

Di Chiro, Giovanna. 1998. Environmental justice from the grassroots: Reflections of history, gender and expertise. In *The Struggle for Ecological Democracy: Environmental Justice Movements in the United States*, ed. Daniel J. Faber, 104–135. New York: Guilford Press.

Dinham, Barbara. 2005. Agrochemical Markets Soar: Pest Pressures or Corporate Design? *Pesticides News* 68 (June):9–11.

DJSI (Dow Jones Sustainability Index). 2009a. <http://www.sustainability-index .com/07_htmle/indexes/overview.html> (accessed September 11, 2009).

DJSI. 2009b. Corporate Sustainability Assessment Criteria. <http://www .sustainability-index.com/07_htmle/assessment/criteria.html> (accessed December 25, 2009).

Dobson, Andrew. 1998. *Justice and the Environment: Conceptions of Environmental Sustainability and Social Justice*. Oxford: Oxford University Press.

Dobson, Andrew. 2003. Social Justice and Environmental Sustainability: Ne'er the Twain Shall Meet? In *Just Sustainabilities: Development in an Unequal World*, ed. Julian Agyeman, Robert D. Bullard, and Bob Evans, 83–95. Cambridge, MA: MIT Press.

Domagalski, Joseph L., Donna L. Knifong, Peter D. Dileanis, Larry R. Brown, Jason T. May, Valerie Connor, and Charles N. Alpers. 2000. Water Quality in the

Sacramento River Basin, California, 1994–98. U.S. Geological Survey Circular 1215. <http://pubs.water.usgs.gov/circ1215> (accessed January 14, 2011).

Donohue, Caitlin. 2009. Out of Reach: How the Sustainable Local Food Movement Neglects Poor Workers and Eaters. *San Francisco Bay Guardian*, December 2. <http://www.sfbg.com/2009/12/02/out-reach> (accessed January 14, 2011).

Dow Accountability Network. 2011. <http://www.thetruthaboutdow.org/index .php> (accessed January 16, 2011).

DPR (Department of Pesticide Regulation). 2001a. Pesticide Drift Minimization Meeting Summary. October 18 (amended December 12).

DPR. 2001b. Regulating Pesticides: The California Story. Department of Pesticide Regulation, Sacramento, CA.

DPR. 2002a. Pesticide Air Monitoring Results Conducted by the Air Resources Board, 1986–2000.

DPR. 2002b. Pesticide Illness Surveillance Program (PISP) Database User Documentation/Dictionary. <http://www.cdpr.ca.gov/docs/whs/pisp.htm> (accessed December 20, 2009).

DPR. 2003. Funding California's Pesticide Regulatory Program: A Report to the Legislature. Pesticide Regulation, Sacramento, CA.

DPR. 2004. DPR Releases Data on 2002 Illnesses. Press Release for 2002 PISP Data. February 26.

DPR. 2006a. Department of Pesticide Regulation Staffing History, 1983/84–2004/05. <http://www.cdpr.ca.gov/docs/dept/budgets/budget.htm> (accessed May 5, 2006).

DPR. 2006b. DPR Releases 2004 Pesticide Use Data; More Nature-Friendly Chemicals Gain Favor. News release. Department of Pesticide Regulation, Sacramento, CA. <http://www.cdpr.ca.gov/docs/pressrls/archive/2006/060124.htm> (accessed January 14, 2011).

DPR. 2006c. Summary of Pesticide Use Report Data 2004, Indexed by Commodity. Department of Pesticide Regulation, Sacramento, CA. <http://www.cdpr .ca.gov/docs/pur/pur04rep/04_pur.htm> (accessed January 14, 2011).

DPR. 2007a. California Pesticide Illness Query System Case #2007–1420–1429. <http://apps.cdpr.ca.gov/calpiq/index.cfm> (accessed January 14, 2011).

DPR. 2007b. Statewide Pesticide Regulatory Activities Summary 2006–2007. <http://www.cdpr.ca.gov/docs/enforce/report5.htm> (accessed July, 8, 2009).

DPR. 2008a. Community Guide to Recognizing and Reporting Pesticide Problems. <http://www.cdpr.ca.gov/docs/dept/comguide/index.htm> (accessed July 9, 2010).

DPR. 2008b. Summary of Ambient Air Monitoring for Methyl Bromide and 1,3-dichloropropene. Memorandum from Shifang Fan to Pamela Wofford. April 21. <http://www.cdpr.ca.gov/docs/emon/pubs/tac/dichlo13.htm> (accessed December 20, 2009).

DPR. 2008c. "Top 100 Pesticides Used Statewide (All Sites Combined) in 2008." <http://www.cdpr.ca.gov/docs/pur/pur08rep/08_pur.htm> (accessed January 14, 2011).

DPR. 2009a. Ambient Air Monitoring for Pesticides in Lompoc, California, Volume 1: Executive Summary. <http://www.cdpr.ca.gov/docs/emon/pubs/tac/tacstdys.htm> (accessed May 26, 2009).

DPR. 2009b. California Statewide Pesticide Regulatory Activities Summaries. <http://www.cdpr.ca.gov/docs/enforce/report5.htm> (accessed July 9, 2009).

DPR. 2009c. County Civil Penalty and Other Administrative Actions. <http://www.cdpr.ca.gov/docs/enforce/admnacts/cvlpnlty.htm> (accessed December 31, 2009).

DPR. 2009d. Environmental Justice Pilot Project Pesticide Air Monitoring in Parlier Addendum ARB Monitoring Results, September 2007. <http://www.cdpr.ca.gov/docs/envjust/pilot_proj> (accessed May 26, 2009).

DPR. 2009e. Final Toxic Air Contaminant (TAC) Evaluation Reports and Determinations. <http://www.cdpr.ca.gov/docs/emon/pubs/tac/finlmenu.htm> (accessed December 30, 2009).

DPR. 2009f. Methyl Iodide External Peer Review Panel. Workshop transcripts. Sacramento, CA, September 25.

DPR. 2009g. Our Budget and Finances. <http://www.cdpr.ca.gov/docs/dept/budgets/budget.htm>. (accessed July 8, 2009).

DPR. 2009h. Our Director and Chief Deputy Director. <http://www.cdpr.ca.gov/dprbios.htm> (accessed December 31, 2009).

DPR. 2009i. Preventing Pesticide Illness: California's Pesticide Illness Surveillance Program. Consumer Fact Sheet. <http://www.cdpr.ca.gov/docs/whs/pisp.htm> (accessed December 20, 2009).

DPR. 2009j. Toxic Air Contaminant Program. <http://www.cdpr.ca.gov/docs/emon/pubs/tacmenu.htm> (accessed September 9, 2009).

DPR. 2009k. Toxic Air Contaminant Program Monitoring Reports. <http://www.cdpr.ca.gov/docs/emon/pubs/tac/tacstdys.htm> (accessed May 26, 2009).

DPR. 2010a. Actively Registered AI's by Common Name. <http://www.cdpr.ca.gov/docs/chemical/monster.htm> (accessed January 1, 2010).

DPR 2010b. Environmental Justice. <http://www.cdpr.ca.gov/docs/envjust> (accessed January 3, 2010).

DPR. 2010c. Evaluation of Chloropicrin as a Toxic Air Contaminant, Part B, Human Health Assessment. Final. California Department of Pesticide Regulation. February 4. <http://www.cdpr.ca.gov/docs/emon/pubs/tac/part_b_0210.pdf> (accessed January 14, 2011).

DPR. 2010d. Memorandum: Director's Proposed Decision concerning Chloropicrin as a Toxic Air Contaminant. April 15. <http://www.cdpr.ca.gov/docs/emon/pubs/tac/finaleval/chloropicrin.htm> (accessed January 14, 2011).

DPR. 2010e. Reducing VOC Emissions from Field Fumigants. <http://www.cdpr.ca.gov/docs/emon/vocs/vocproj/reg_fumigant.htm> (accessed January 1, 2010).

Dubrovsky, Neil M., Charles R. Kratzer, Larry R. Brown, JoAnn M. Gronberg, and Karen R. Burow. 1998. Water Quality in the San Joaquin-Tulare Basins, California,

1992–95. U.S. Geological Survey Circular 1159. <http://water.usgs.gov/pubs/circ /circ1159> (accessed January 14, 2011).

Duhigg, Charles. 2009. Debating How Much Weed Killer Is Safe in Your Water Glass. *New York Times*, August 23, A1.

DuPont. 2009. <http://www2.dupont.com/Production_Agriculture/en_US/cpp _us.html> (accessed November 10, 2009).

DuPuis, E. Melanie. 2002. *Nature's Perfect Food: How Milk Became America's Drink*. New York: New York University Press.

DuPuis, E. Melanie, ed. 2004. *Smoke and Mirrors: The Politics and Culture of Air Pollution*. New York: New York University Press.

DuPuis, E. Melanie. 2007. Angels and Vegetables: A Brief History of Food Advice in America. *Gastronomica* 7 (2): 34–44.

DuPuis, E. Melanie, and David Goodman. 2005. Should We Go "Home" to Eat? Toward a Reflexive Politics of Localism. *Journal of Rural Studies* 21 (3):359–371.

DuPuis, E. Melanie, David Goodman, and Jill Harrison. 2007. Just Values or Just Value? Remaking the Local in Agro-Food Studies. In *Between the Local and the Global: Confronting Complexity in the Contemporary Agri-Food Sector*, ed. Terry Marsden and Jonathan Murdoch, 241–268. Oxford: Elsevier.

DuPuis, E. Melanie, Jill Lindsey Harrison, and David Goodman. 2011. Just Food? In *Cultivating Food Justice*, ed. Alison Hope Alkon and Julian Agyeman. Cambridge, MA: MIT Press.

Egilman, David S., and Marion A. Billings. 2005. Abuse of Epidemiology: Automobile Manufacturers Manufacture a Defense to Asbestos Liability. *International Journal of Occupational and Environmental Health* 11 (4):360–371.

Epstein, Lynn, and Susan Bassein. 2003. Patterns of Pesticide Use in California and the Implications for Strategies for Reduction of Pesticides. *Annual Review of Phytopathology* 41: 23.1–23.25.

Eschbach, Karl, Jacqueline Hagan, Nestor Rodriguez, Ruben Hernandez-Leon, and Stanley Bailey. 1999. Death at the Border. *International Migration Review* 33 (2): 430–454.

Estabrook, Barry. 2009. Politics of the Plate: The Price of Tomatoes. *Gourmet*, March. <http://www.gourmet.com/magazine/2000s/2009/03/politics-of-the-plate -the-price-of-tomatoes?currentPage=1> (accessed November 20, 2009).

ETC Group. 2005. Oligopoly, Inc. *Communique* 91 (November/December 2005). <http://www.etcgroup.org/en/materials/publications> (accessed January 14, 2011).

European Union. 2006. Draft Assessment Report for Chloropicrin, Volume 1, May. Ministry of Health, Italy. Available at <http://dar.efsa.europa.eu/dar-web /provision> (accessed January 14, 2011).

European Union. 2011. Registration, Evaluation, Authorisation, and Restriction of Chemicals. <http://ec.europa.eu/enterprise/sectors/chemicals/reach/index_en.htm> (accessed January 16, 2011).

Faber, Daniel, ed. 1998. *The Struggle for Ecological Democracy: Environmental Justice Movements in the United States*. New York: Guilford Press.

Faber, Daniel. 2008. *Capitalizing on Environmental Injustice: The Polluter-Industrial Complex in the Age of Globalization.* Lanham, MD: Rowman and Littlefield.

Faber, Daniel, and Debra McCarthy. 2003. Neoliberalism, Globalization, and the Struggle for Ecological Democracy: Linking Sustainability and Environmental Justice. In *Just Sustainabilities: Development in an Unequal World,* ed. Julian Agyeman, Robert D. Bullard, and Bob Evans, 38–63. Cambridge, MA: MIT Press.

Fan, Anna. 2002. Revised Findings on the Health Effects of Methyl Isothiocyanate. Memorandum from Office of Environmental Health Hazard Assessment to Department of Pesticide Regulation. January 31.

Fan, Anna. 2003. Comments on the Risk Characterization Document for Inhalation Exposure to Methyl Bromide, Addendum to Volume I, Prepared by the Department of Pesticide Regulation. Office of Environmental Health Hazard Assessment. March 11.

Farm Worker Pesticide Project. 2006. Poisons on the Wind: Community Air Monitoring for Chlorpyrifos in the Yakima Valley. Report by Farm Worker Pesticide Project and Pesticide Action Network North America. <http://www.fwpp.org/?page=FWPPReports> (accessed January 14, 2011).

Feldman, Jay, et al. 2009. Letter from Beyond Pesticides and 29 other co-signees to Lisa P. Jackson, U.S. EPA Administrator, regarding EPA's May 27, 2009, fumigant cluster decision. June 18, 2009.

Figueroa, Robert M. 2003. Bivalent Environmental Justice and the Culture of Poverty. *Rutgers University Journal of Law and Urban Policy* 1 (1): 27–42.

Fitzgerald, Deborah Kay. 2003. *Every Farm a Factory: The Industrial Ideal in American Agriculture.* New Haven, CT: Yale University Press.

FitzSimmons, Margaret. 1986. The New Industrial Agriculture: The Regional Integration of Specialty Crop Production. *Economic Geography* 62 (4): 334–353.

Foucault, Michel. (1978) 1990. The History of Sexuality, Volume 1: An Introduction. New York: Vintage Books.

Fraser, Nancy. 2000. Rethinking Recognition. *New Left Review* 3 (May/June):107–120.

Freidberg, Suzanne. 2004. *French Beans and Food Scares: Culture and Commerce in an Anxious Age.* New York: Oxford University Press.

Freidberg, Suzanne. 2009. *Fresh: A Perishable History.* Cambridge, MA: Belknap Press.

Friedland, William H., Amy E. Barton, and Robert J. Thomas. 1981. *Manufacturing Green Gold: Capital, Labor, and Technology in the Lettuce Industry.* Cambridge: Cambridge University Press.

Friedland, William H., Lawrence Busch, Frederick H. Buttel, and Alan P. Rudy. 1991. *Towards a New Political Economy of Agriculture.* Boulder, CO: Westview Press.

Friedman, Milton. 1970. The Social Responsibility of Business is to Increase Its Profits. *New York Times Magazine,* September 13: 32–33, 122–126.

Friedmann, Harriet. 2007. Scaling Up: Bringing Public Institutions and Food Service Corporations into the Project for a Local, Sustainable Food System in Ontario. *Agriculture and Human Values* 24 (3): 389–398.

Froines, John. 2010. Report of the Scientific Review Committee on Methyl Iodide to the Department of Pesticide Regulation. February 5. <http://www.cdpr.ca.gov /docs/risk/mei/peer_review_report.pdf> (accessed January 14, 2011).

Fujii, Reed. 2009. Farmers Focus on Safe Spraying. *Stockton Record*, July 16.

Galarza, Ernesto. 1964. *Merchants of Labor: The Mexican Bracero Story*. Santa Barbara, CA: McNally and Loftin.

Galt, Ryan E. 2008. Beyond the Circle of Poison: Significant Shifts in the Global Pesticide Complex, 1976–2008. *Global Environmental Change* 18 (4): 786–799.

GAO (General Accounting Office). 2001. Information on Pesticide Illness Reporting Systems. Statement of John B. Stephenson, Director, Natural Resources and Environment Team of GAO. GAO-01-501T.

GAO. 2005. Environmental Justice: EPA Should Devote More Attention to Environmental Justice When Developing Clean Air Rules. Report GAO-05-289.

GAO. 2006. Special Report to Senate Majority Leader Frist on Illegal Immigration. Report GAO-06-077.

Garcia, Natalie. 2006. Pesticides Help Farmers But Can Hurt Neighbors. *Visalia Times-Delta*, August 19.

Gelobter, Michel. 2002. Foreward to *Toxic Struggles: The Theory and Practice of Environmental Justice*, ed. Richard Hofrichter. Salt Lake City: University of Utah Press.

Gennaro, Valerio, and Lorenzo Tomatis. 2005. Business Bias: How Epidemiologic Studies May Underestimate or Fail to Detect Increased Risks of Cancer and Other Diseases. *International Journal of Occupational and Environmental Health* 11 (4):356–359.

Getches, David H., and David N. Pellow. 2002. Beyond 'Traditional' Environmental Justice. In *Justice and Natural Resources: Concepts, Strategies, and Applications*, ed. Kathryn M. Mutz, Gary C. Bryner, and Douglas S. Kenney, 3–30. Washington, DC: Island Press.

Getz, Christy, Sandy Brown, and Aimee Shreck. 2008. Class Politics and Agricultural Exceptionalism in California's Organic Agriculture Movement. *Politics and Society* 36 (4): 478–507.

Gibbs, Lois. 1982. *Love Canal: My Story*. Albany: State University of New York Press.

Gieryn, Thomas F. 1999. *Cultural Boundaries of Science: Credibility on the Line*. Chicago: University of Chicago Press.

Gliessman, Stephen R. 2007. *Agroecology: The Ecology of Sustainable Food Systems*. 2nd ed. Boca Raton: CRC Press.

Glotfelty, D. E., J. N. Seiber, and L. A. Liljedahl. 1987. Pesticides in Fog. *Nature* 325:602–605.

Goldman, Patti, Janette K. Brimmer, and Virginia Ruiz. 2009. Pesticides in the Air—Kids at Risk: Petition to EPA to Protect Children from Pesticide Drift. <http:// earthjustice.org/our_work/campaigns/pesticides-in-the-air-kids-at-risk> (accessed January 14, 2011).

Goodman, David, Bernardo Sorj, and John Wilkinson. 1987. *From Farming to Biotechnology: A Theory of Agro-Industrial Development*. New York: Basil Blackwell.

Gottlieb, Robert. 1988. *A Life of Its Own: The Politics and Power of Water*. San Diego: Harcourt Brace Jovanovich.

Gottlieb, Robert, ed. 1995. *Reducing Toxics: A New Approach to Policy and Industrial Decisionmaking*. Washington, DC: Island Press.

Gottlieb, Robert. 2001. *Environmentalism Unbound: Exploring New Pathways for Change*. Cambridge, MA: MIT Press.

Gottlieb, Robert. 2005. *Forcing the Spring: The Transformation of the American Environmental Movement*. Rev. ed. Washington, DC: Island Press.

Gottlieb, Robert, and Andrew Fisher. 1996a. Community Food Security and Environmental Justice: Searching for a Common Discourse. *Agriculture and Human Values* 3 (3): 23–32.

Gottlieb, Robert, and Andrew Fisher. 1996b. "First Feed the Face": Environmental Justice and Community Food Security. *Antipode* 28 (2): 193–203.

Gottlieb, Robert, and Anupama Joshi. 2010. *Food Justice*. Cambridge, MA: MIT Press.

Gould, Kenneth A., David N. Pellow, and Allan Schnaiberg. 2004. Interrogating the Treadmill of Production: Everything You Wanted to Know But Were Afraid to Ask. *Organization and Development* 17 (3): 296–316.

Guha, Ramachandra. 1998. Radical American Environmentalism: A Third World Critique. In *The Great New Wilderness Debate*, ed. J. Baird Callicott and Michael P. Nelson, 231–245. Athens: University of Georgia Press.

Guillette, Elizabeth A., Maria M. Meza, Maria G. Aquilar, Alma D. Soto, and Idalia E. Garcia. 1998. An Anthropological Approach to the Evaluation of Preschool Children Exposed to Pesticides in Mexico. *Environmental Health Perspectives* 106 (6): 347–353.

Guthman, Julie. 2004. *Agrarian Dreams: The Paradox of Organic Farming in California*. Berkeley: University of California Press.

Guthman, Julie. 2007a. Can't Stomach It: How Michael Pollan et al. Made Me Want to Eat Cheetos. *Gastronomica* 7 (2): 75–79.

Guthman, Julie. 2007b. Commentary on Teaching Food: Why I Am Fed Up with Michael Pollan et al. *Agriculture and Human Values* 24 (2): 261–264.

Guthman, Julie. 2008a. Bringing Good Food to Others: Investigating the Subjects of Alternative Food Practice. *Cultural Geographies* 15 (4): 431–447.

Guthman, Julie. 2008b. "If Only They Knew": Color Blindness and Univeralism in California Alternative Food Institutions. *Professional Geographer* 60 (3): 387–397.

Guthman, Julie. 2008c. Neoliberalism and the Making of Food Politics in California. *Geoforum* 39 (3): 1171–1183.

Guthman, Julie. 2008d. Thinking Inside the Neoliberal Box: The Micro-Politics of Agro-Food Philanthropy. *Geoforum* 39 (3): 1241–1253.

Harris, Gardiner, and Robert Pear. 2007. Ex-Officials Tell of Tension over Science and Politics. *New York Times*, July 12.

Harrison, Jill. 2006. "Accidents" and Invisibilities: Scaled Discourse and the Naturalization of Regulatory Neglect in California's Pesticide Drift Conflict. *Political Geography* 25 (5): 506–529.

Harrison, Jill. 2008. Abandoned Bodies and Spaces of Sacrifice: Pesticide Drift Activism and the Contestation of Neoliberal Environmental Politics in California. *Geoforum* 39 (3): 1197–1214.

Harrison, Jill Lindsey. 2011. Parsing "Participation" in Action Research: Navigating the Challenges of Lay Involvement in Technically Complex Participatory Science Projects. *Society and Natural Resources* 24 (7).

Harvey, David. 1996. *Justice, Nature, and the Geography of Difference*. Malden, MA: Blackwell.

Harvey, David. 2005. *A Brief History of Neoliberalism*. New York: Oxford University Press.

Hassenein, Neva. 1999. *Changing the Way America Farms: Knowledge and Community in the Sustainable Agriculture Movement*. Lincoln: University of Nebraska Press.

Hayes, Tyrone B. 2004. There Is No Denying This: Defusing the Confusion about Atrazine. *Bioscience* 54 (12): 1138–1149.

Hayes, Tyrone B., Kelly Haston, Mable Tsui, Anhthu Hoang, Cathryn Haeffele, and Aaron Vonk. 2002. Herbicides: Feminization of Male Frogs in the Wild. *Nature* 419 (6910): 895–896.

Heavner, Brad. 1999. Toxics on Tap: Pesticides in California Drinking Water Sources. Californians for Pesticide Reform and California Public Research Group Charitable Trust, San Francisco.

Hedberg, Carl-Johan, and Fredrik von Malmborg. 2003. The Global Reporting Initiative and Corporate Sustainability Reporting in Swedish Companies. *Corporate Social Responsibility and Environmental Management* 10 (3): 153–164.

Heffernan, William H. 2000. Concentration of Ownership and Control in Agriculture. In *Hungry for Profit: The Agribusiness Threat to Farmers, Food, and the Environment*, ed. Fred Magdoff, John Bellamy Foster, and Frederick H. Buttel, 61–75. New York: Monthly Review Press.

Heindel, Jerrold J., Robert E. Chapin, Dushyant K. Gulati, Julia D. George, Catherine J. Price, Melissa C. Marr, Christina B. Myers, Leta H. Barnes, Patricia A. Fail, Thomas B. Grizzle, Bernard A. Schwetz, and Raymond S. H. Yang. 1994. Assessment of the Reproductive and Developmental Toxicity of Pesticide/Fertilizer Mixtures Based on Confirmed Pesticide Contamination in California and Iowa Groundwater. *Fundamental and Applied Toxicology* 22 (4): 605–621.

Henderson, George L. 1999. *California and the Fictions of Capital*. New York: Oxford University Press.

Henke, Christopher. 2008. *Cultivating Science, Harvesting Power: Science and Industrial Agriculture in California*. Cambridge, MA: MIT Press.

Hightower, Jim. 1973. *Hard Tomatoes, Hard Times*. Cambridge, MA: Schenkman.

Hinrichs, C. Clare. 2003. The Practice and Politics of Food System Localization. *Journal of Rural Studies* 19 (1): 33–45.

Hinrichs, C. Clare, and Thomas A. Lyson. 2007. *Remaking the North American Food System: Strategies for Sustainability*. Lincoln: University of Nebraska Press.

Hofrichter, Richard. 1993. *Toxic Struggles: The Theory and Practice of Environmental Justice*. Philadelphia: New Society Publishers.

Holifield, Ryan. 2004. Neoliberalism and Environmental Justice in the United States Environmental Protection Agency: Translating Policy into Managerial Practice in Hazardous Waste Remediation. *Geoforum* 35 (3): 285–297.

Holifield, Ryan, Michael Porter, and Gordon Walker. 2009. Introduction: Spaces of Environmental Justice: Frameworks for Critical Engagement. *Antipode* 41 (4): 591–612.

Holmes, Seth M. 2007. "Oaxacans Like to Work Bent Over": The Naturalization of Social Suffering among Berry Farm Workers. *International Migration* 45 (3): 39–68.

Howard, Philip H. 2009a. Visualizing Consolidation in the Global Seed Industry: 1996–2008. *Sustainability* 1 (4): 1266–1287.

Howard, Philip H. 2009b. Visualizing Food System Concentration and Consolidation. *Southern Rural Sociology* 24 (2): 87–110.

Hsu, Eric. 2003a. Residents at Risk: Drifts Worry Residents. *Bakersfield Californian*, November 10.

Hsu, Eric. 2003b. Severity of Pesticide Drift Fuels Worry. *Bakersfield Californian*, October 9.

Hunold, Christian, and Iris Marion Young. 1998. Justice, Democracy, and Hazardous Siting. *Political Studies* 46 (1): 82–95.

Infante, Peter F. 2005. Safeguarding Scientific Evaluations by Governmental Agencies: Case Study of OSHA and the 1,3-butadiene Classification. *Journal of Occupational and Environmental Health* 11 (4): 372–377.

International Coalition for Justice in Bhopal. 2011. <http://bhopal.net/oldsite/icjb.html> (accessed January 16, 2011).

Irwin, Alan. 2001. *Sociology and the Environment: A Critical Introduction to Society, Nature, and Knowledge*. Malden, MA: Blackwell.

Irwin, Alan, Henry Rothstein, Steven Yearly, and Elaine McCarthy. 1997. Regulatory Science: Towards a Sociological Framework. *Futures* 29 (1): 17–31.

Jackson, Lisa. 2009. Remarks to the National Environmental Justice Advisory Council. July 21. <http://yosemite.epa.gov/opa/admpress.nsf/8d49f7ad4bbcf4ef852573590040b7f6/313ec9a2bc80d677852575fa007b3c42!OpenDocument> (accessed January 14, 2011).

Jacobson, Michael F. 2005. Lifing the Veil of Secrecy from Industry Funding of Nonprofit Health Organizations. *International Journal of Occupational and Environmental Health* 11 (4): 349–355.

Janofsky, Michael. 2006. Unions Say EPA Bends to Political Pressure. *New York Times*, August 2.

Jasanoff, Sheila. 1990. *The Fifth Branch: Science Advisors as Policmakers*. Cambridge, MA: Harvard University Press.

Jasanoff, Sheila. 1992. Science, Politics, and the Renegotiation of Expertise at EPA. *Osiris* 7: 195–217.

Jasanoff, Sheila. 2003. A Living Legacy: The Precautionary Ideal in American Law. In *Precaution: Environmental Science and Preventive Public Policy*, ed. Joel Tickner, 227–240. Washington, DC: Island Press.

Jelinik, Lawrence J. 1979. *Harvest Empire: A History of California Agriculture*. San Francisco: Boyd and Fraser.

Jessop, Bob. 2002. Neoliberalism and Urban Governance: A State Theoretical Perspective. *Antipode* 34 (3): 452–472.

Johansen, Erik W. 2010. Withdrawl of Application for Registration of Pesticides in Washington. Letter from Erik W. Johansen, Special Pesticide Registration Program Coordinator for the State of Washington, Department of Agriculture, to Rodney C. Akers. Arysta LifeScience North America, LLC. July 16.

Jones, Katherine T. 1998. Scale as Epistemology. *Political Geography* 17 (1):25–28.

Kegley, Susan, and David Chatfield. 2004. Letter to Paul Helliker, Director of the DPR. May 27.

Kegley, Susan, Anne Katten, and Marion Moses. 2003. Secondhand Pesticides: Airborne Pesticide Drift in California. Pesticide Action Network, San Francisco.

Kegley, Susan, Karl A. Tupper, Katherine T. Mills, Forest M. Kaser, Elizabeth Stover, Andrew C.-K. Wang, Amy M. Wu, Carol Dansereau, Sara Bjorkvist, Stephanie Jim, Tanya Brown, Christine Reardon, Chela Vazquez, and Jacqueline J. Epps. 2009. Community-Based Environmental Monitoring of Airborne Pesticides: Volatilization Drift of Semi-Volatile Pesticides as a Contributor to Bystander Pesticide Exposure. Draft paper. November 25.

Kern County. 2009. County of Kern: Agriculture and Measurement Standards. <http://www.kernag.com/default.asp> (accessed December 31, 2009).

Keystone Alliance. 2009. Field to Market: The Keystone Alliance for Sustainable Agriculture. Environmental Resource Indicators for Measuring Outcomes of On-Farm Agricultural Production in the United States. Executive Summary <http://www.keystone.org/spp/environment/sustainability/field-to-market> (accessed September 11, 2009).

Khokha, Sasha. 2009. Reporter's Notes: Catching the Drift. KQED Radio. October 16. <http://www.kqed.org/quest/radio/catching-the-drift> (accessed October 31, 2009).

King, Andrew, and Michael Lenox. 2000. Industry Self-regulation without Sanctions: The Chemical Industry's Responsible Care Program. *Academy of Management Journal* 43 (4): 698–716.

King, Leslie, and Deborah McCarthy. 2009. *Environmental Sociology: From Analysis to Action*. Lanham: Rowman and Littlefield.

Kjaergard, Lise L., and Bodil Als-Nielsen. 2002. Association between Competing Interests and Authors' Conclusions: Epidemiological Study of Randomised Clinical Trials Published in the British Medical Journal. *British Medical Journal* 325 (7358): 249.

Kleinman, Daniel Lee. 2003. *Impure Cultures: University Biology and the World of Commerce*. Madison: University of Wisconsin Press.

Kloppenburg, Jack Ralph. 2005. *First the Seed: The Political Economy of Plant Biotechnology*. 2nd ed. Madison: University of Wisconsin Press.

Kloppenburg, Jack, Jr., and Neva Hassanein. 2006. From Old School to Reform School? *Agriculture and Human Values* 23 (4): 417–421.

Kloppenburg, Jack Ralph, John Henrickson, and G. W. Stevenson. 1996. Coming into the Foodshed. *Agriculture and Human Values* 13 (3): 33–42.

Kosek, Jake. 2004. Purity and Pollution: Racial Degradation and Environmental Anxieties. In *Liberation Ecologies: Environment, Development, Social Movements*, ed. Richard Peet and Michael Watts, 125–165. London: Routledge.

Kovarik, William. 2005. Ethyl-Leaded Gasoline: How a Classic Occupational Disease Becomes an International Public Health Disaster. *International Journal of Occupational and Environmental Health* 11 (4): 384–397.

Krimsky, Sheldon. 2003. *Science in the Private Interest: Has the Lure of Profits Corrupted Biomedical Research?* Lanham, MD: Rowman and Littlefield Publishers.

Kroma, Margaret, and Cornelia Butler Flora. 2003. Greening Pesticides: A Historical Analysis of the Social Construction of Farm Chemical Advertisements. *Agriculture and Human Values* 20 (1): 21–35.

Kuletz, Valerie L. 1998. *The Tainted Desert: Environmental Ruin in the American West*. New York: Routledge.

Kurtz, Hilda E. 2003. Scale Frames and Counter-Scale Frames: Constructing the Problem of Environmental Injustice. *Political Geography* 22 (8):887–916.

Kymlicka, Will. 2002. *Contemporary Political Philosophy: An Introduction*. 2nd ed. Oxford: Oxford University Press.

Lake, Robert W. 1996. Volunteers, NIMBYs, and Environmental Justice: Dilemmas of Democratic Practice. *Antipode* 28 (2): 160–174.

Land Stewardship Project and PAN. 2010. The Syngenta Corporation and Atrazine: The Cost to the Land, People, and Democracy. Land Stewardship Project and Pesticide Action Network. <http://www.landstewardshipproject.org/pdf /AtrazineReportJan2010.pdf> (accessed January 14, 2011).

Langston, Nancy. 2010. *Toxic Bodies: Hormone Disruptors and the Legacy of DES*. New Haven, CT: Yale University Press.

Lappé, Frances Moore, and Anna Lappé. 2002. *Hope's Edge: The Next Diet for a Small Planet*. New York: Jeremy P. Tarcher/Putnam.

Lardner, George, Jr., and Joby Warrick. 2000. Pesticide Coalition Tries to Blunt Regulation. *Washington Post*, May 13, A01.

Larson, Alice C. 2000. An Assessment of Worker Training under the Worker Protection Standard: Executive Summary. Prepared for the U.S. EPA Office of Pesticide Programs. <http://www.epa.gov/oppfead1/safety/newnote/exec_summary .pdf> (accessed January 14, 2011).

Lee, Charles, ed. 2002. *Proceedings: The First National People of Color Environmental Leadership Summit.* New York: United Church of Christ Commission for Racial Justice.

Lee, Sharon, Robert McLaughlin, Martha Harnly, Robert Gunier, and Richard Kreutzer. 2002. Community Exposures to Airborne Agricultural Pesticides in California: Ranking of Inhalation Risks. *Environmental Health Perspectives* 110 (12): 1175–1184.

Leitner, Helga Jamie Peck, and Eric S. Sheppard, eds. 2007. *Contesting Neoliberalism: Urban Frontiers.* New York: Guilford.

LeNoir, James S., Laura L. McConnell, Gary M. Fellers, Thomas M. Cahill, and James N. Seiber. 1999. Summertime Transport of Current-Use Pesticides from California's Central Valley to the Sierra Nevada Mountain Range, USA. *Environmental Toxicology and Chemistry* 18 (2): 2715–2722.

Levine, Marvin J. 2007. *Pesticides: A Toxic Time Bomb in Our Midst.* Westport, CT: Praeger.

Littlefield, Amy. 2009. A Drifting Danger for Central Valley Schoolchildren. *Los Angeles Times*, August 17.

Lobao, Linda, and Katherine Meyer. 2001. The Great Agricultural Transition: Crisis, Change, and Social Consequences of Twentieth Century Farming. *Annual Review of Sociology* 27: 103–124.

London, Jonathan, Julie Sze, and Raoul S. Lievanos. 2008. Problems, Promise, Progress, and Perils: Critical Reflections on Environmental Justice Policy Implementation in California. *UCLA Journal of Environmental Law and Policy* 26 (2): 255–289.

Low, Nicholas, and Brendan Gleeson. 1998. *Justice, Society, and Nature: An Exploration of Political Ecology.* London: Routledge.

Lucier, Gary, and Rachel L. Dettman, 2008. Vegetables and Melons Outlook. USDA Economic Research Service. June 26.

Lukes, Steven. 2006. *Individualism.* Colchester, UK: ECPR Press.

Lyson, Thomas A. 2004. *Civic Agriculture: Reconnecting Farm, Food, and Community.* Medford, MA: Tufts University Press.

Lyson, Thomas A., G. W. Stevenson, and Rick Welsh, eds. 2008. *Food and the Mid-Level Farm: Renewing an Agriculture of the Middle.* Cambridge, MA: MIT Press.

Magdoff, Fred, John Bellamy Foster, and Frederick H. Buttel, eds. 2000. *Hungry for Profit: The Agribusiness Threat to Farmers, Food, and the Environment.* New York: Monthly Review Press.

Majka, Linda C., and Theo J. Majka. 2000. Organizing U.S. Farm Workers: A Continuous Struggle. In *Hungry for Profit: The Agribusiness Threat to Farmers, Food, and the Environment*. Ed. Fred Magdoff, John Bellamy Foster, and Frederick H. Buttel, 161-174. New York: Monthly Review Press.

Marquardt, Sandra, Caroline Cox, and Holly Knight. 1998. Toxic Secrets: "Inert" Ingredients in Pesticides, 1987–1997. Californians for Pesticide Reform and Northwest Coalition for Alternatives to Pesticides, San Francisco.

Martinez, Robert, Terrance B. Gratton, Claudia Coggin, Antonio Rene, and William Waller. 2004. A Study of Pesticide Safety and Health Perceptions among Pesticide Applicators in Tarrant County, Texas. *Journal of Environmental Health* 66 (6): 34–37.

Massey, Douglas. 2002. *Beyond Smoke and Mirrors: Mexican Immigration in an Era of Economic Integration*. New York: Russell Sage Foundation.

Maxwell, Lesli. 1999. Earlimart Pesticide Case Baffles Officials: Application That Made Dozens Ill Appears Legal, Inspectors Say. *Fresno Bee*, November 16, A1.

Maxwell, Lesli. 2000. Aid for Pesticide Ailments Sought. *Fresno Bee*, February 16, B1.

McConnell, Grant. 1966. *Private Power and American Democracy*. New York: Knopf.

McGarity, Thomas O. 2001. Politics by Other Means: Law, Science, and Policy in US EPA's Implementation of the Food Quality Protection Act. *Administrative Law Review* 53 (1): 103–222.

McMichael, Philip. 1994. *The Global Restructuring of Agro-Food Systems*. Ithaca, NY: Cornell University Press.

McWilliams, Carey. [1939] 1999. *Factories in the Field: The Story of Migratory Farm Labor in California*. Berkeley: University of California Press.

McWilliams, Carey. 1973. *Southern California: An Island on the Land*. Layton, UT: Gibbs Smith.

Merchant, Carolyn. 2003. Shades of Darkness: Race and Environmental History. *Environmental History* 8 (3): 380–394.

Meuter, Michael. 2008. Comments to US EPA on the Fumigant Cluster Assessment Based on Experience at Moss Landing, California. California Rural Legal Assistance, Inc. October 30. No longer available online.

Mitchell, Don. 1996. *The Lie of the Land: Migrant Workers and the California Landscape*. Minneapolis: University of Minnesota Press.

Mitchell, Don. 2001. The Devil's Arm: Points of Passage, Networks of Violence, and the California Agricultural Landscape. *New Formations* 43:44–60.

Mitchell, Don. 2003. *The Right to the City: Social Justice and the Fight for Public Space*. New York: Guilford Press.

Mitchell, Don. 2007. Work, Struggle, Death, and Geographies of Justice: The Transformation of Landscape in and beyond California's Imperial Valley. *Landscape Research* 32 (5): 559–577.

Mohai, Paul, David Pellow, and J. Timmons Roberts. 2009. Environmental Justice. *Annual Review of Environment and Resources* 34: 405–430.

Montague, Peter. 2000. Foreword. *Making Better Environmental Decisions: An Alternative to Risk Assessment*, ed. Mary O'Brien. Cambridge, MA: MIT Press.

Moore, Monica, and Steve Scholl-Buckwald. n.d. Note from the Directors: Accelerating Reform. Pesticide Action Network. No longer available online.

Morello-Frosch, Rachel, Stephen Zavestoski, Phil Brown, Rebecca Gasior Altman, Sabrina McCormick, and Brian Mayer. 2006. Embodied Health Movements: Responses to a "Scientized" World. In *The New Political Sociology of Science: Institutions, Networks, and Power*, ed. Scott Frickel and Kelly Moore. Madison: University of Wisconsin Press.

Morris, Amy Wilson. 2008. Easing Conservation? Conservation Easements, Public Accountability, and Neoliberalism. *Geoforum* 39 (3): 1215–1227.

Moses, Marion. 1993. Farmworkers and Pesticides. In *Confronting Environmental Racism*, ed. Robert D. Bullard, 161–178. Boston: South End Press.

Moses, Marion, Eric S. Johnson, W. Kent Anger, Virlyn W. Burse, Sandford W. Horstman, Richard J. Jackson, Robert G. Lewis, Keith T. Maddy, Rob McConnell, William J. Meggs, and Sheila Hoar Zahm. 1993. Environmental Equity and Pesticide Exposure. *Toxicology and Industrial Health* 9 (5): 913–959.

Mullenix, Phyllis. 2005. Fluoride Poisoning: A Puzzle with Hidden Pieces. *International Journal of Occupational and Environmental Health* 11 (4): 404–414.

Murray, Douglas, Catharina Wesseling, Matthew Keifer, Marianela Corriols, and Samuel Henao. 2002. Surveillance of Pesticide-Related Illness in the Developing World: Putting the Data to Work. *Journal of International Occupational and Environmental Health* 8 (3): 243–248.

Muzzling Those Pesky Scientists. 2006. Editorial. *New York Times*, December 11.

Nash, Linda. 2006. *Inescapable Ecologies: A History of Environment, Disease, and Knowledge*. Berkeley: University of California Press.

National Agricultural Aviation Association. 2009. Professional Aerial Applicator Support System. <http://www.agaviation.org/paass.htm> (accessed October 28, 2009).

National Agricultural Workers Survey. 2002. Demographic and Employment Profile of United States Farm Workers. <http://www.doleta.gov/agworker/naws.cfm>.

National Coalition on Drift Minimization. 2009. <http://pep.wsu.edu/ncodm> (accessed December 25, 2009).

National Institute for Occupational Safety and Health. 2009. National Institute for Occupational Safety and Health Pesticide Illness and Injury Surveillance. <http://www.cdc.gov/niosh/topics/pesticides> (accessed December 2, 2009).

Nestle, Marion. 2002. *Food Politics: How the Food Industry Influences Nutrition and Health*. Berkeley: University of California Press.

Nestle, Marion. 2006. *What to Eat*. New York: North Point.

Nevins, Joseph. 2002. *Operation Gatekeeper: The Rise of the "Illegal Alien" and the Making of the U.S.-Mexico Boundary*. New York: Routledge.

Nevins, Joseph. 2005. A Beating Worse Than Death: Imagining and Contesting Violence in the U.S.-Mexico Borderlands. *AmeriQuests* 2 (1).

Nevins, Joseph. 2008. *Dying to Live: A Story of U.S. Immigration in an Age of Global Apartheid*. San Francisco: City Lights.

Northwest Coalition for Alternatives to Pesticides. 2009. Pesticide Use Reporting Program. <http://www.pesticide.org> (accessed November 22, 2009).

Nozick, Robert. 1974. *Anarchy, State, and Utopia*. New York: Basic Books.

NRC (National Research Council). 1989. *Alternative Agriculture*. Washington, DC: National Academy Press.

NRC. 2009. *Science and Decisions: Advancing Risk Assessment*. Washington, DC: National Academy Press.

NRDC. 2003. Complaint for Declaratory and Injunctive Relief. Lawsuit against US EPA by Natural Resources Defense Council and Other Plaintiffs. September 15.

Nussbaum, Martha. 2005. Beyond the Social Contract: Capabilities and Global Justice. In *The Political Philosophy of Cosmopolitanism*, ed. Gillian Brock, 196–218. Cambridge: Cambridge University Press.

O'Brien, Mary. 2000. *Making Better Environmental Decisions: An Alternative to Risk Assessment*. Cambridge, MA: MIT Press.

Occupational Safety and Health Administration. 1998. Informational Booklet on Industrial Hygiene. U.S. Department of Labor. <http://www.osha.gov/Publications /OSHA3143/OSHA3143.htm> (accessed July 28, 2010).

Olvera, Javier Erik. 1999a. Earlimart Crowd Seeking Answers: Residents Quiz a Panel about the Lasting Effects of Saturday's Pesticide Scare. *Fresno Bee*, November 18, A16.

Olvera, Javier Erik. 1999b. Earlimart Residents Urge Review of Pesticide Policies. *Fresno Bee*, December 9, A20.

O'Malley, Michael. 2004. Pesticides. In *Current Occupational and Environmental Medicine*, ed. Joseph LaDou. New York: Lange Medical Books.

Organic Farming Research Foundation. 2010. About OFRF. <http://ofrf.org/about us/aboutus.html> (accessed March 2, 2010).

Organic Trade Association. 2007. Organic Trade Association 2007 Manufacturer Survey. <http://www.ota.com/organic/mt/business.html> (accessed January 2, 2010).

O'Rourke, Dara, and Gregg P. Macey. 2003. Community Environmental Policing: Assessing New Strategies of Public Participation in Environmental Regulation. *Journal of Policy Analysis and Management* 22 (3): 383–414.

Ottinger, Gwen E. 2010. Buckets of Resistance: Standards and the Effectiveness of Citizen Science. *Science, Technology, and Human Values* 35 (March): 244–270.

PAN (Pesticide Action Network). 2006. Air Monitoring for Chlorpyrifos in Lindsay, California: June–July 2004 and July–August 2005. <http://www.panna.org>.

PAN. 2007. Air Monitoring for Pesticides in Hastings, Florida, December 6–14, 2006: Technical Report. <http://www.panna.org>.

PAN. 2008. Air Monitoring in Hastings, Florida, October 1–December 6, 2007: Technical Report. <http://www.panna.org>.

PAN. 2009a. "DOJ Begins Monsanto Probe." PAN Update. October 15.

PAN. 2009b. Drift Catcher Results. <http://www.panna.org>.

PAN. 2009c. Endosulfan. <http://www.panna.org> (accessed December 30, 2009).

PAN. 2009d. "EPA to Screen 67 Pesticides for Endocrine Risks." *PAN North America Magazine* (Summer), 3.

PAN. 2009e. PANNA Corporate Profiles. <http://www.panna.org> (accessed August 6, 2009).

PAN. 2009f. "Tell Obama's Antitrust Czar: Investigate Monsanto." <http://action.panna.org/t/5185/petition.jsp?petition_KEY=2096> (accessed November 13, 2009).

PAN. 2010a. Monsanto and CropLife Men Have No Place in Government. <http://action.panna.org/t/5185/petition.jsp?petition_KEY=2150> (accessed January 1, 2010).

PAN. 2010b. Pesticide Contributions to the Ozone Problem in California. <http://www.panna.org> (accessed January 1, 2010).

PAN. 2010c. PAN letter to DPR Director Mary-Ann Warmerdam. Comments on DPR's proposed decision to register methyl iodide. June 29, 2010.

PAN. 2011. Safe Strawberries. <http://www.panna.org/issues/related-umbrella -campaign/cancer-free-strawberries> (accessed January 14, 2011).

Panousis, Jim. 2008. Ag's Largest Advertisers. *AgriMarketing* 46 (6): 33.

Patel, Rajeev. 2007. Transgressing Rights: La Via Campesina's Call for Food Sovereignty. *Feminist Economics* 13 (1): 87–101.

Patel, Rajeev, Robert J. Torres, and Peter Rosset. 2005. Genetic Engineering in Agriculture and Corporate Engineering in Public Debate: Risk, Public Relations, and Public Debate over Genetically Modified Crops. *International Journal of Occupational and Environmental Health* 11 (4): 428–436.

Pearce, Fred, and Debora Mackenzie. 1999. It's Raining Pesticides: The Water Falling from Our Skies Is Unfit to Drink. *New Scientist* 162 (2180): 23.

Pease, William S., Rachel A. Morello-Frosch, David S. Albright, Amy D. Kyle, and James C. Robinson. 1993. Preventing Pesticide-Related Illness in California Agriculture: Strategies and Priorities. Policy seminar, Berkeley, CA.

Peck, Jamie. 2004. Geography and Public Policy: Constructions of Neoliberalism. *Progress in Human Geography* 28 (3): 392–405.

Peck, Jamie, and Adam Tickell. 2002. Neoliberalizing Space. *Antipode* 34 (3): 380–404.

PEER (Public Employees for Environmental Responsibility). 2006a. EPA Begins Closing Libraries before Congress Acts on Plan. News Release. August 21. <http://www.peer.org/news/news_id.php?row_id=731> (accessed January 14, 2011).

PEER. 2006b. EPA Scientists Protest Pending Pesticide Approvals. News Release. May 25. <http://www.peer.org/news/news_id.php?row_id=691> (accessed January 14, 2011).

PEER. 2006c. Taking the "E" out of EPA. News release. May 8. <http://www.peer.org/news/news_id.php?row_id=685> (accessed January 14, 2011).

Pelletier, David, Vivica Kraak, Christine McCullum, Ulla Uusital, and Robert Rich. 1999. The Shaping of Collective Values through Deliberative Democracy: An Empirical Study from New York's North Country. *Policy Sciences* 32 (2): 103–131.

Pellow, David Naguib. 2000. Environmental Inequality Formation: Toward a Theory of Environmental Injustice. *American Behavioral Scientist* 43 (4): 581–601.

Pellow, David Naguib. 2005. *Garbage Wars: The Struggle for Environmental Justice in Chicago.* Cambridge, MA: MIT Press.

Pellow, David Naguib, and Robert J. Brulle, eds. 2005. *Power, Justice, and the Environment: A Critical Appraisal of the Environmental Justice Movement.* Cambridge, MA: MIT Press.

Peña, Devon. 2005. Autonomy, Equity, and Environmental Justice. In *Power, Justice, and the Environment: A Critical Appraisal of the Environmental Justice Movement,* ed. David N. Pellow and Robert J. Brulle, 131–151. Cambridge, MA: MIT Press.

Perrow, Charles. 1984. *Normal Accidents: Living with High-Risk Technologies.* New York: Basic Books.

Perry, Melissa, and Frederick R. Bloom. 1998. Perceptions of Pesticide Associated Cancer Risks among Farmers: A Qualitative Assessment. *Human Organization* 57 (3): 342–349.

Pesticide Drift Handled Properly, Panel Says. 2004. *Los Angeles Times,* March 4, B7.

Pimentel, David, and Lois Levitan. 1986. Pesticides: Amounts Applied and Amounts Reaching Pests. *Bioscience* 36: 86–91.

Pincetl, Stephanie S. 1999. *Transforming California: A Political History of Land Use and Development.* Baltimore: Johns Hopkins University Press.

PISP (Pesticide Illness Surveillance Program). 1999. Pesticide Illness Surveillance Program 1999 Summary Report. <http://www.cdpr.ca.gov/docs/whs/pisp.htm> (accessed January 15, 2011).

PISP. 2000. Pesticide Illness Surveillance Program 2000 Summary Report. <http://www.cdpr.ca.gov/docs/whs/pisp.htm> (accessed January 15, 2011).

PISP. 2002. Pesticide Illness Surveillance Program 2002 Summary Report. <http://www.cdpr.ca.gov/docs/whs/pisp.htm> (accessed January 15, 2011).

PISP. 2003. Pesticide Illness Surveillance Program 2003 Summary Report. <http://www.cdpr.ca.gov/docs/whs/pisp.htm> (accessed January 15, 2011).

PISP. 2004. Pesticide Illness Surveillance Program 2004 Summary Report. <http://www.cdpr.ca.gov/docs/whs/pisp.htm> (accessed January 15, 2011).

PISP. 2005. Pesticide Illness Surveillance Program 2005 Summary Report. <http://www.cdpr.ca.gov/docs/whs/pisp.htm> (accessed January 15, 2011).

PISP. 2006. Pesticide Illness Surveillance Program 2006 Summary Report. <http://www.cdpr.ca.gov/docs/whs/pisp.htm> (accessed January 15, 2011).

PISP. 2007. Pesticide Illness Surveillance Program 2007 Summary Report. <http:// www.cdpr.ca.gov/docs/whs/pisp.htm> (accessed January 15, 2011).

Players Discuss Current System for Pesticide Regulation. 2005. *Bakersfield Californian*, September 17.

Pollan, Michael. 2006. *The Omnivore's Dilemma: A Natural History of Four Meals*. New York: Penguin.

Pollan, Michael. 2008. *In Defense of Food: An Eater's Manifesto*. New York: Penguin.

Poppendieck, Janet. 1998. *Sweet Charity? Emergency Food and the End of Entitlement*. New York: Viking.

Poppendieck, Janet. 2000. Want Amid Plenty: From Hunger to Inequality. In *Hungry for Profit: The Agribusiness Threat to Farmers, Food, and the Environment*, ed. Fred Magdoff, John Bellamy Foster, and Frederick H. Buttel, 189–202. New York: Monthly Review Press.

Porter, Warren P., James W. Jaeger, and Ian H. Carlson. 1999. Endocrine, Immune, and Behavioral Effects of Aldicarb (Carbamate), Atrazine (Triazine), and Nitrate (Fertilizer) Mixtures at Groundwater Concentrations. *Toxicology and Industrial Health* 15 (1–2): 133–151.

Prado, Celia, and Don Villarejo. 1996. Farmworker WPS Training. Unpublished study. California Institute for Rural Studies, Davis.

Pudup, Mary Beth. 2008. It Takes a Garden: Cultivating Citizen-Subjects in Organized Garden Projects. *Geoforum* 39 (3): 1228–1240.

Pulido, Laura. 1996a. A Critical Review of the Methodology of Environmental Racism Research. *Antipode* 28 (2): 142–159.

Pulido, Laura. 1996b. *Environmentalism and Economic Justice*. Tucson: University of Arizona Press.

Purcell, Mark, and Joseph Nevins. 2005. Pushing the Boundary: State Restructuring, State Theory, and the Case of U.S.-Mexico Border Enforcement in the 1990s. *Political Geography* 24 (2): 211–235.

Quandt, Sara A., Thomas A. Arcury, Colin K. Austin, and Rosa M. Saavedra. 1998. Farmworker and Farmer Perceptions of Farmworker Agricultural Chemical Exposure in North Carolina. *Human Organization* 57 (3): 359–368.

Raffensperger, Carolyn, and Joel Tickner, eds. 1999. *Protecting Public Health and the Environment: Implementing the Precautionary Principle*. Washington, DC: Island Press.

Raju, Manu. 2005. EPA's Draft Equity Plan Drops Race as a Factor in Decisions. *Inside EPA*, July 1. <http://www.precaution.org/lib/06/prn_epa_drops_race_from _ej_guide.050701.htm> (accessed December 21, 2009).

Rampton, Sheldon, and John Stauber. 2002. *Trust Us, We're Experts! How Industry Manipulates Science and Gambles with Your Future*. New York: Jeremy P. Tarcher.

Rankin Bohme, Susanna, John Zorabedian, and David S. Egilman. 2005. Maximizing Profit and Endangering Health: Corporate Strategies to Avoid Litigation and Regulation. *International Journal of Occupational and Environmental Health* 11 (4): 338–348.

Rao, Pamela, Thomas A. Arcury, Sara Quandt, and Alicia Doran. 2004. North Carolina Growers' and Extension Agents' Perceptions of Latino Farmworker Pesticide Exposure. *Human Organization* 63 (2): 151–161.

Ray, Daryll E., Daniel G. De La Torre Ugarte, and Kelly Tiller. 2003. Rethinking US Agricultural Policy: Changing Course to Secure Farmer Livelihoods Worldwide. Knoxville: Agricultural Policy Analysis Center, University of Tennessee. <http://www.agpolicy.org/blueprint.html> (accessed January 15, 2011).

Rees, Joseph. 1997. Development of Communitarian Regulation in the Chemical Industry. *Law and Policy* 19 (4): 477–528.

Reeves, Margaret, Anne Katten, and Martha Guzmán. 2002. Fields of Poison 2002: California Farmworkers and Pesticides. Pesticide Action Network, San Francisco.

Reisner, Marc. 1993. *Cadillac Desert: The American West and Its Disappearing Water.* Penguin.

Responsible Care. 2009. <http://www.responsiblecare.org/page.asp?p=6401&l=1> (accessed September 30, 2009).

Ritter, John. 2005. In California's Central Valley, Pesticide Fight Heats Up. *USA Today*, April 11.

Robbins, Paul. 2004. *Political Ecology: A Critical Introduction.* Malden, MA: Blackwell.

Robbins, Paul. 2007. *Lawn People: How Grasses, Weeds, and Chemicals Make Us Who We Are.* Philadelphia: Temple University Press.

Roberts, Eric M., Paul B. English, Judith K. Grether, Gayle C. Windham, Lucia Somberg, and Craig Wolff. 2007. Maternal Residence Near Agricultural Pesticide Applications and Autism Spectrum Disorders among Children in the California Central Valley. *Environmental Health Perspectives* 115 (10): 1482–1489.

Rodriguez, Robert. 2005. Raisin Farmer Wins $7.5 Million Verdict. *Fresno Bee*, April 15.

Rohr, Jason R., and Krista A. McCoy. 2010. A Qualitative Meta-analysis Reveals Consistent Effects of Atrazine on Freshwater Fish and Amphibians. *Environmental Health Perspectives* 118 (1): 20–32.

Roosevelt, Margot. 2009. California Farmland Fumigants Challenged. *Los Angeles Times*, May 26.

Ross, Zev. 1998. Toxic Fraud: Deceptive Advertising by Pest Control Companies in California. Californians for Pesticide Reform and California Public Interest Research Group. <http://pesticidereform.org/article.php?id=13> (accessed January 1, 2010).

Rowe, Gene, and Lynn J. Frewer. 2005. A Typology of Public Engagement Mechanisms. *Science, Technology, and Human Values* 30 (2): 251–290.

Rudy, Alan P. 2006. Neoliberalism, Neoconservativism, and the Spaces of and for Coalition. *Agriculture and Human Values* 23 (4): 423–425.

Russell, Edmund. 2001. *War and Nature: Fighting Humans and Insects with Chemicals from World War I to Silent Spring.* Cambridge: Cambridge University Press.

Sabina, Ellen. 2010. Proposed Pesticide Use in California Presents Another Reason to Buy Local Food. Just Means Sustainable Food Blog. June 26. <http://www .justmeans.com/Proposed-Pesticide-Use-in-California-Presents-Another-Reason -Buy-Local-Food/20423.html> (accessed January 15, 2011).

Sanborn, Margaret, Donald Cole, Kathleen Kerr, Cathy Vakil, Luz Helena Sanin, and Kate Bassil. 2004. Pesticides Literature Review. Ontario College of Family Physicians, Toronto.

Sandler, Ronald, and Phaedra C. Pezzullo. 2007. *Environmental Justice and Environmentalism: The Social Justice Challenge to the Environmental Movement.* Cambridge, MA: MIT Press.

Sandweiss, Stephen. 1998. *Environmental Justice and the New Pluralism: The Challenge of Difference in Environmentalism.* Oxford: Oxford University Press.

Sass, Jennifer Beth. 2005. Industry Efforts to Weaken the EPA's Classification of the Carcinogenicity of 1,3-butadiene. *International Journal of Occupational and Environmental Health* 11 (4): 378–383.

Sass, Jennifer Beth, and Mae Wu. 2007. Budget Cuts to the U.S. EPA Will Reduce Government Data on Pollutants, and Increase Reliance on Industry Data. *International Journal of Occupational and Environmental Health* 13 (2): 244–246.

Schlosberg, David. 2007. *Defining Environmental Justice: Theories, Movements, and Nature.* Oxford: Oxford University Press.

Schlosser, Eric. 2001. *Fast Food Nation: The Dark Side of the All-American Meal.* New York: Houghton Mifflin.

Schuette, J. D., J. Weaver, J. Troiano, and J. Dias. 2002. Update of the Well Inventory Database, California Department of Pesticide Regulation. <http://www.cdpr .ca.gov/docs/empm/pubs/ehapreps.htm> (accessed January 15, 2011).

Sen, Amartya. 1993. Capability and Well-Being. In *The Quality of Life*, ed. Amartya Sen and Martha Nussbaum, 30–53. Oxford: Clarendon Press.

Serafini, Maureen P. 2009. Withdrawal of Applications to Register the New Active Ingredient (NAI) Iodomethane. Letter from Maureen P. Serafini, Director of Bureau of Pesticides Management, New York State Department of Environmental Conservation, to Teresa McMullin. Arysta LifeScience North America LLC. January 14.

Shah, Sonia. 2010. Behind Mass Die-offs, Pesticides Lurk as Culprit. *Yale 360*, January 7. <http://www.e360.yale.edu/content/feature.msp?id=2228> (accessed January 15, 2011).

Shilling, Fraser M., Jonathan K. London, and Raoul S. Lievanos. 2009. Marginalization by Collaboration: Environmental Justice as a Third Party in and beyond CALFED. *Environmental Science & Policy* 12 (6): 694–709.

Shipp, Eva M., Sharon P. Cooper, Deborah J. del Junco, Jane N. Bolin, Ryan E. Whitworth, and Charles J. Cooper. 2007. Pesticide Safety Training among Farmworker Adolescents from Starr County, Texas. *Journal of Agricultural Safety and Health* 13 (3): 311–321.

Shrader-Frechette, Kristin. 2002. *Environmental Justice: Creating Equality, Reclaiming Democracy*. Oxford: Oxford University Press.

Shreck, Aimee, Christy Getz, and Gail Feenstra. 2006. Social Sustainability, Farm Labor, and Organic Agriculture: Findings from an Exploratory Analysis. *Agriculture and Human Values* 23 (4): 439–449.

Sifry, Micah, and Nancy Watzman. 2004. *Is That a Politician in Your Pocket? Washington on $2 Million a Day*. Hoboken, NJ: Wiley.

Singer, Peter, and Jim Mason. 2006. *The Way We Eat: Why Our Food Choices Matter*. New York: Rodale.

Slocum, Rachel. 2007. Whiteness, Space, and Alternative Food Practice. *Geoforum* 38 (3):520–533.

Smith, Neil. 1984. *Uneven Development: Nature, Capital, and the Production of Space*. Oxford: Blackwell.

Snow, David A., and Robert D. Benford. 1988. Ideology, Frame Resonance, and Participant Mobilization. *International Social Movement Research* 1: 197–218.

Snow, David A., and Robert D. Benford. 1992. Master Frames and Cycles of Protest. In *Frontiers in Social Movement Theory*, ed. Aldon D. Morris and Carol M. Mueller. New Haven, CT: Yale University Press.

Snow, David A., E. Burke Rochford, Steven K. Worden, and Robert D. Benford. 1986. Frame Alignment Processes, Micromobilization, and Movement Participation. *American Sociological Review* 51 (4): 464–481.

Snyder, Caroline. 2005. The Dirty Work of Promoting "Recycling" of America's Sewage Sludge. *International Journal of Occupational and Environmental Health* 11 (4):415–427.

Source Watch. 2009. Center for Media Democracy. <http://www.sourcewatch.org /index.php/Government-industry_revolving_door> (accessed December 26, 2009).

Spitzer, Skip. 2005. A Systematic Approach to Occupational and Environmental Health. *International Journal of Occupational and Environmental Health* 11 (4): 444–455.

Spitzer, Skip. n.d. Industrial Agriculture and Corporate Power. Pesticide Action Network. <http://www.panna.org> (accessed November 13, 2009).

Spray Safe. 2011. <http://www.foodandfarming.info/spraysafe.asp> (accessed January 4, 2011).

Stapleton, Lisa K. 2003. Catching Drift. *Terrain* 34 (Summer):18–22. <http:// ecologycenter.org/terrain/issues/summer-2003/catching-drift> (accessed January 15, 2011).

Starr, Kevin. 1985. *Inventing the Dream: California through the Progressive Era*. New York: Oxford University Press.

Stauber, John C., and Sheldon Rampton. 1995. *Toxic Sludge Is Good for You! Lies, Damn Lies, and the Public Relations Industry*. Monroe, ME: Common Courage Press.

Stephen, Lynn. 2004. The Gaze of Surveillance in the Lives of Mexican Immigrant Workers. *Development* 47 (1): 97–102.

Stoll, Steven. 1998. *The Fruits of Natural Advantage: Making the Industrial Countryside in California.* Berkeley: University of California Press.

Strange, Marty. 1988. *Family Farming: A New Economic Vision.* Lincoln: University of Nebraska Press.

Strochlic, Ron, Cathy Wirth, Ana Fernandez Besada, and Christy Getz. 2008. Farm Labor Conditions on Organic Farms in California. California Institute for Rural Studies, Davis, CA.

Szasz, Andrew. 1994. *Ecopopulism: Toxic Waste and the Movement for Environmental Justice.* Minneapolis: University of Minnesota Press.

Szasz, Andrew. 2008. *Shopping Our Way to Safety: How We Changed from Protecting the Environment to Protecting Ourselves.* Minneapolis: University of Minnesota Press.

Sze, Julie. 2007. *Noxious New York: The Racial Politics of Urban Health and Environmental Justice.* Cambridge, MA: MIT Press.

Sze, Julie, Jonathan London, Fraser Shilling, Gerardo Gambirazzio, Trina Filan, and Mary Cadenasso. 2009. Defining and Contesting Environmental Justice: Socio-Natures and the Politics of Scale in the Delta. *Antipode* 41 (4): 807–843.

Sunstein, Cass R. 2002. *Risk and Reason: Safety, Law, and the Environment.* Cambridge: Cambridge University Press.

Tarr, Joel. 1996. *The Search for the Ultimate Sink: Urban Pollution in Historical Perspective.* Akron, OH: University of Akron Press.

Taylor, Dorceta. 2000. The Rise of the Environmental Justice Paradigm: Injustice Framing and the Social Construction of Environmental Discourses. *American Behavioral Scientist* 43 (4): 508–580.

Taylor, Dorceta. 2009. *The Environment and the People in American Cities, 1600s–1900s: Disorder, Inequality, and Social Change.* Durham, NC: Duke University Press.

Taylor, Jennifer. 2004. *Earlimart: Pesticides in the Valley.* Documentary. Aired on California Connected, June 17. < <http://www.californiaconnected.org/tv/archives/category/story-topic/environment> (accessed January 15, 2011).

Thelin, Gail P., and Leonard P. Gianessi. 2000. Methods for Estimating Pesticide Use for County Areas of the Conterminous United States. U.S Geological Survey Open-File Report 2000-25. <http://water.usgs.gov/nawqa/pnsp/pubs/ofr00250> (accessed January 16, 2011).

Thomas, Robert J. 1985. *Citizenship, Gender, and Work.* Berkeley: University of California Press.

Thornton, Joe. 2000. *Pandora's Poison: Chlorine, Health, and a New Environmental Strategy.* Cambridge, MA: MIT Press.

Tickner, Joel A. 2003. *Precaution, Environmental Science, and Preventive Public Policy.* Washington, DC: Island Press.

Tupper, Karl, Susan Kegley, and Brian Hill. 2008. PAN Comments to US EPA on the Fumigant Reregistration Eligibility Decision. October 30. <http://www.panna.org>.

Unhealthy Influence. 2007. Editorial. *New York Times*, July 12.

Union of Concerned Scientists. 2009. Chemical Industry Pressures EPA to Protect Herbicide, Not Wildlife. <http://www.ucsusa.org/scientific_integrity/abuses_of_science/atrazine-and-health.html> (accessed December 31, 2009).

United Nations Environment Program. 2004. Childhood Pesticide Poisoning: Information for Advocacy and Action. Geneva: United Nations Environment Program. <http://www.who.int/ceh/publications/pestipoison/en> (accessed January 15, 2011).

U.S. Department of Food and Agriculture. 2002. California State Level Data. In vol. 1, *Census of Agriculture*. <http://www.agcensus.usda.gov/Publications/2002/Census_by_State/California/index.asp> (accessed January 15, 2011).

U.S. Department of Agriculture. 2011. USDA National Organic Program. <http://www.ams.usda.gov/AMSv1.0/NOP> (accessed January 4, 2011).

U.S. EPA (Environmental Protection Agency). 2004. Pesticides Industry Sales and Usage 2000 and 2001 Market Estimates. Office of Prevention, Pesticides, and Toxic Substances. <http://www.epa.gov/opp00001/pestsales> (accessed January 15, 2011).

U.S. EPA. 2006. Evaluation Report: EPA Needs to Conduct Environmental Justice Reviews of Its Programs, Policies, and Activities. U.S. EPA Office of Inspector General. Report No. 2006-P-00034. September 18. <http://www.epa.gov/oig/reports/2006/20060918-2006-P-00034.pdf> (accessed January 15, 2011).

U.S. EPA. 2008a. Field Volatilization of Agricultural Pesticides. Pesticide Program Dialogue Committee Meeting Presentation Slides. May 21. <http://www.epa.gov/pesticides/ppdc/2008/may2008/may08.htm> (accessed January 15, 2011).

U.S. EPA. 2008b. FY 2008/2009 OPP Budget Update. Office of Pesticide Programs. October 7. <http://www.epa.gov/pesticides/ppdc/2008/oct2008/session-4.pdf> (accessed January 15, 2011).

U.S. EPA. 2008c. Reregistration Eligibility Decision (RED) for Methyl Bromide. July 9. <http://www.epa.gov/opp00001/reregistration/methyl_bromide> (accessed January 15, 2011).

U.S. EPA. 2009b. Chloropicrin: Final Revised HED Human Health Risk Assessment. U.S. Environmental Protection Agency, Office of Pesticide Programs, Health Effects Division (7509C), May. <http://www.regulations.gov> (under docket ID No. EPA-HQ-OPP-2007–0350–0396) (accessed January 15, 2011).

U.S. EPA. 2009c. Extension of Conditional Registration of Iodomethane (Methyl Iodide). <http://www.epa.gov/opp00001/factsheets/iodomethane_fs.htm> (accessed December 30, 2009).

U.S. EPA. 2009d. Pesticides: Regulating Pesticides. <http://www.epa.gov/pesticides/regulating/index.htm> (accessed June 9, 2009).

U.S. EPA. 2009e. Revised Risk Assessment Methods for Workers, Children of Workers in Agricultural Fields, and Pesticides with No Food Uses. Office of Pesticide Programs. December 7. <http://www.epa.gov/pesticides/health/revised RAmethods.pdf> (accessed January 15, 2011).

U.S. EPA. 2009f. Statement from Steve Owens, Assistant Administrator for the Office of Prevention, Pesticides, and Toxic Substances (OPPTS). <http://www.epa .gov/pesticides/regulating/statement-owens.html> (accessed December 31, 2009).

U.S. EPA. 2009g. Table 3.6: Most Commonly Used Conventional Pesticide Active Ingredients in the U.S. Agricultural Market Sector. <http://www.epa.gov /opp00001/pestsales/01pestsales/usage2001_2.htm> (accessed June 22, 2009).

U.S. EPA. 2009h. 2009 Stratospheric Ozone Protection Awards. <http://www.epa .gov/Ozone/awards/winners_2009.html> (accessed July 27, 2010).

U.S. EPA. 2010a. Food Quality Protection Act (FQPA) of 1996. <http://www.epa .gov/pesticides/regulating/laws/fqpa> (accessed July 24, 2010).

U.S. EPA. 2010b. Transmittal of Meeting Minutes of the FIFRA SAP Meeting Held December 1–3, 2009 on the Scientific Issues Associated with "Field Volatilization of Conventional Pesticides." Office of Prevention, Pesticides, and Toxic Substances. February 25. Arlington, VA.

U.S. Geological Survey. 2011. U.S. Geological Survey Pesticide National Synthesis Project. <http://water.usgs.gov/nawqa/pnsp> (accessed January 8, 2011).

U.S. Office of Management and Budget. 2010. Budget of the United States Government. Budget Authority for U.S. Environmental Protection Agency.

Vallas, Steven P., and Daniel Lee Kleinman. 2008. Contradiction, Convergence, and the Knowledge Economy: The Co-Evolution of Academic and Commercial Biotechnology. *Socio-Economic Review* 6 (2): 283–311.

Van den Bosch, Robert. 1978. *The Pesticide Conspiracy*. Garden City, NY: Doubleday.

Van Steenwyk, Robert, and Frank G. Zalom. 2005. Environmental Laws Elicit Evolution in Pest Management. *California Agriculture* 59 (1):2.

Van Tassell, Larry W., Mark A. Ferrell, Bozheng Yang, David E. Legg, and John E. Lloyd. 1999. Pesticide Practices and Perceptions of Wyoming Farmers and Ranchers. *Journal of Soil and Water Conservation* 54 (1): 410–415.

Varsanyi, Monica. 2008a. Immigration Policing through the Backdoor: City Ordinances, the "Right to the City," and the Exclusion of Undocumented Day Laborers. *Urban Geography* 29 (1): 29–52.

Varsanyi, Monica. 2008b. Should Cops be La Migra? *Los Angeles Times*, April 20.

Vasquez, Raul. 2005. Trouble in the Air: Dozens Feel Ill after Toxic Gas Drifts into Neighborhood. *Monterey County Weekly*, October 13.

Vaughan, Elaine. 1995. The Socioeconomic Context of Exposure and Response to Environmental Risk. *Environment and Behavior* 27 (4): 454–489.

Via Campesina. 2011. What Is Food Sovereignty? <http://viacampesina.org /en/index.php?option=com_content&view=article&id=47:food-sovereignty &catid=21:food-sovereignty-and-trade&Itemid=38> (accessed January 15, 2011).

Villarejo, Don, David Lighthall, Daniel Williams III, Ann Souter, Richard Mines, Bonnie Bade, Steve Samuels, and Stephen A. McCurdy. 2000. Suffering in Silence: A Report on the Health of California's Agricultural Workers. California Institute for Rural Studies, Davis, CA.

Villarejo, Don, and Stephen A. McCurdy. 2008. The California Agricultural Workers Health Survey. *Journal of Agricultural Safety and Health* 14 (2): 135–146.

Vos, Timothy. 2000. Visions of the Middle Landscape: Organic Farming and the Politics of Nature. *Agriculture and Human Values* 17 (3): 245–256.

Walker, Richard. 2004. *The Conquest of Bread: 150 Years of Agribusiness in California*. New York: New Press.

Warner, Keith. 2007. *Agroecology in Action: Extending Alternative Agriculture through Social Networks*. Cambridge, MA: MIT Press.

Washburn, Jennifer. 2005. *University Inc: The Corporate Corruption of Higher Education*. New York: Basic Books.

Wedel, Janine R. 2009. *Shadow Elite: How the World's New Power Brokers Undermine Democracy, Government, and the Free Market*. New York: Basic Books.

Weinberg, Justine Lew, Lisa J. Bunin, and Rupali Das. 2009. Application of the Industrial Hygiene Hierarchy of Controls to Prioritize and Promote Safer Methods of Pest Control: A Case Study. *Public Health Reports* 124 (supplement 1): 53–62.

Weir, David, and Mark Schapiro. 1981. *Circle of Poison: Pesticides and People in a Hungry World*. San Francisco: Institute for Food and Development Policy.

Welch, Dwight, Steve Shapiro, Dave Christenson, Mark Coryell, Larry Penley, Wendell Smith, Patrick Chan, John O'Grady, and Paul Scoggins. 2006. Letter to US EPA Administrator Stephen Johnson from Local Presidents of EPA Unions Representing Scientists, Risk Managers, and Related Staff. May 24. <http://www .panna.org/resources/fqpa> (accessed December 30, 2009).

Wells, Miriam. 1996. *Strawberry Fields: Politics, Class, and Work in California Agriculture*. Ithaca, NY: Cornell University Press.

Wenz, Peter S. 1988. *Environmental Justice*. Albany: State University of New York Press.

White, Richard. 1995. *The Organic Machine: The Remaking of the Columbia River*. New York: Hill and Wang.

Whiteside, Kerry H. 2006. *Precautionary Politics: Principle and Practice in Confronting Environmental Risk*. Cambridge, MA: MIT Press.

Whitney, Susan P. 2000. Measuring Adoption of New Drift-Reduction Practices among Pesticide Applicators as a Result of Training. University of Delaware Cooperative Extension System, Newark. <http://ag.udel.edu/EXTENSION/pesticide /behavior_adoption/drift.htm> (accessed November 6, 2009).

Whorton, James. 1974. *Before Silent Spring: Pesticides and Public Health in Pre-DDT America*. Princeton, NJ: Princeton University Press.

Williams, Raymond. 1973. *The Country and the City*. New York: Oxford University Press.

Williams, Raymond. 1980. *Problems in Materialism and Culture: Selected Essays*. London: New Left Books.

Wilson, Geoff. 2001. From Productivism to Post-Productivism . . . and Back Again? Exploring the (Un)changed Natural and Mental Landscapes of European Agriculture. *Transactions of the Institute of British Geographers* 26 (1): 77–102.

Winders, Bill. 2009. *The Politics of Food Supply: U.S. Agricultural Policy in the World Economy.* New Haven, CT: Yale University Press.

Winders, Jamie. 2007. Bringing Back the (B)order: Post-9/11 Politics of Immigration, Borders, and Belonging in the Contemporary U.S. South. *Antipode* 39 (5): 920–942.

Wing, Steve. 2000. Limits of Epidemiology. In *Illness and the Environment: A Reader in Contested Medicine,* ed. Steve Kroll-Smith, Phil Brown, and Valerie J. Gunter, 29–45. New York: New York University Press.

Wingspread. 1998. Wingspread Statement on the Precautionary Principle. <http://www.gdrc.org/u-gov/precaution-3.html> (accessed January 15, 2011).

Winne, Mark. 2008. *Closing the Food Gap: Resetting the Table in the Land of Plenty.* Boston: Beacon.

Wright, Wynne, and Gerad Middendorf. 2008. *The Fight Over Food: Producers, Consumers, and Activists Challenge the Global Food System.* University Park: University of Pennsylvania Press.

Wu, Mae, Mayra Quirindongo, Jennifer Sass, and Andrew Wetzler. 2009. Poisoning the Well: How the EPA Is Ignoring Atrazine Contamination in Surface and Drinking Water in the Central United States. Natural Resources Defense Council, New York.

Young, Iris Marion. 1983. Justice and Hazardous Waste. *Bowling Green Studies in Applied Philosophy* 5: 171–183.

Young, Iris Marion. 1990. *Justice and the Politics of Difference.* Princeton, NJ: Princeton University Press.

Zabik, John M., and James N. Sieber. 1993. Atmospheric Transport of Organophosphate Pesticides from California's Central Valley to the Sierra Nevada Mountains. *Journal of Environmental Quality* 22 (1): 80–90.

Zalom, Frank G., Nick C. Toscano, and Frank J. Byrne. 2005. Managing Resistance Is Critical to Future Use of Pyrethroids and Neonicotinoids. *California Agriculture* 59 (1): 11–15.

Zamora, Celia, Charles R. Kratzer, Michael S. Majewski, and Donna L. Knifong. 2003. Diazinon and Chlorpyrifos Loads in Precipitation and Urban and Agricultural Storm Runoff during January and February 2001 in the San Joaquin River Basin, California. U.S. Geological Survey, Water-Resources Investigations Report 03-4091. <http://pubs.usgs.gov/wri/wri034091> (accessed January 15, 2011).

Index